Electronics for Absolute Beginne

Electronics for Absolute Beginners

Published by Philip Dixon

Publication Date and Edition: January 2021, 2nd Edition. Reprinted: February 2021.

Limit of Liability/Disclaimer of Warranty

Preface

***Chambers Twentieth Century Dictionary* (1972) defines electronics as "The science and technology of the conduction of electricity in a vacuum, a gas, or a semiconductor, and devices based thereon".**

Today, electronic devices are everywhere ... computers, cell (mobile) phones, televisions, calculators, hearing aids, washing machines, and cars are just a few examples of devices that form part of everyday life in the 21st century and which make use of electronics. Without electronics our lives would be very different indeed.

Early breakthroughs, discoveries and inventions that have paved the way to where we are today began in the late 18th century, with many scientists, engineers and mathematicians making significant contributions over the years. People such as the Italian physicist Allesandro Volta (1745-1827), who invented the battery in 1800, German physicist Georg Simon Ohm (1789-1854), who defined the relationship between current, voltage and resistance, and French physicist and mathematician André-Marie Ampère (1775-1836), who studied the effects of electric current and invented the solenoid. Other famous names include English scientist Michael Faraday (1791-1867) who demonstrated electromagnetic induction, Irish physicist George Johnstone Stoney (1826-1911) who suggested that electricity must be "built" out of tiny electrical charges and later coined the name "electron", and Scottish scientist James Clerk Maxwell (1831-1879) who carried out significant work on magnetism, electromagnetism and electricity, and formulated the electromagnetic theory of light.

The actual history of electronics began with the invention in 1904 of the vacuum diode by English electrical engineer and physicist John Ambrose Fleming (1849-1945), which quickly led to the invention by American electrical engineer Lee De Forest (1873-1961) of the first electronic amplifying device – the triode. This in turn led to the tetrode and pentode tubes that were dominant until World War II.

During the early part of the 20th century, progress was made in certain technical and scientific fields that led, in 1947, to the invention of the transistor by American physicists John Bardeen (1908-1991), William Shockley (1910-1989) and Walter Houser Brattain (1902-1987). Of all the electronic components that exist today, the transistor is the one that has had the biggest impact on electronics and the computer industry in general. The transistor is the building block of modern electronic systems. Without transistors, many of the mobile devices we use today would not be mobile at all: laptops, mobile phones, and tablets would be too big to fit in a room in your house, let alone be *portable*.

The study of electronics makes for a fantastic hobby because it provides a great mix of mathematics and science, and gives you the opportunity to design and build your own gadgets. Let your imagination run wild and enjoy the amazing world of electronics.

About this Book

The purpose of this book is to introduce the subject of electronics to people who have no prior knowledge of the subject. If you don't know what a prototyping breadboard is, or have no idea what a resistor or diode do, then this book will help you to gain a basic understanding of what electronics components exist and what they are used for. You will also learn how to combine these components to build electronics circuits.

We don't cover anything to do with mains voltage.

How this Book is Organized

Although there is a logical flow in the way that information is presented, you can pretty much jump around from chapter to chapter as it suits you, picking out the material that is of most interest.

The book is split into two parts – the first introduces the subject of electronics, and the second contains a range of simple circuits to design and build.

Part 1 – Getting Started with Electronics

- Chapter 1 - Electricity and Electric Circuits
- Chapter 2 - Equipment You Need in Your Electronics Laboratory
- Chapter 3 - Safety
- Chapter 4 - Resistors
- Chapter 5 - Capacitors
- Chapter 6 - Diodes
- Chapter 7 - Light Emitting Diodes (LEDs)
- Chapter 8 - Transistors
- Chapter 9 - Integrated Circuits
- Chapter 10 - The 555 Timer
- Chapter 11 - Combining Electronics with Software

Part 2 - Designing and Building Electronics Circuits

- Chapter 12 - Using an Electronics Prototyping Breadboard
- Chapter 13 - Illuminating a Single LED
- Chapter 14 - Using a Multimeter to Measure Voltage, Current and Resistance
- Chapter 15 - Connecting Multiple LEDs in Series
- Chapter 16 - Connecting Multiple LEDs in Parallel
- Chapter 17 - Using a Variable Resistor to Control the Speed of a Small DC Motor
- Chapter 18 - Using Wind Power

- Chapter 19 - Using a Transistor to Switch On an LED
- Chapter 20 - Charging and Discharging a Capacitor
- Chapter 21 - Building a Light Sensor - Night Light
- Chapter 22 - Introduction to Solar Power

Quick Start Workbook Website

The Quick Start Workbook website contains information about all the books available from the author. You can find the site here:

http://www.quickstartworkbook.com/

YouTube Videos

All the videos referenced in this book are available on the Quick Start Workbook YouTube channel. You can find the channel here:

https://www.youtube.com/channel/UChoBHeUk6tc6Si2hrdOYJOw

Table of Contents

Part 1 - Getting Started with Electronics

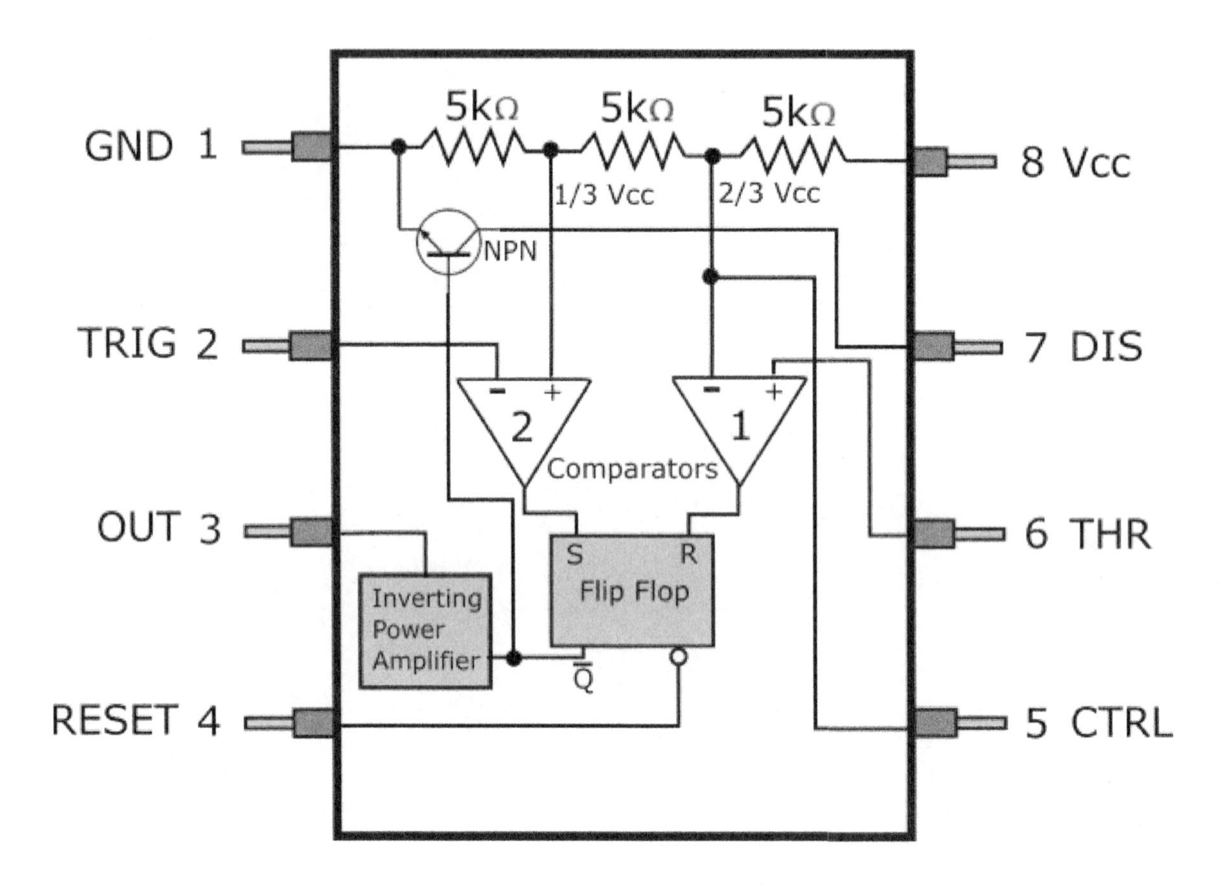

Chapter 1

Electricity and Electric Circuits

In order to get the most out of this book and to have a better chance of understanding subsequent chapters, it's a good idea to learn a bit about electricity and electric circuits before we go any further.

Atoms, Protons, Neutrons, and Electrons

All matter (any substance that has mass and takes up space) is made up of tiny particles called atoms. Copper is made up of copper atoms, oxygen is made up of oxygen atoms, and so on. An atom is the smallest particle of a particular substance that can exist. For example, if you split a copper atom into two parts, those two parts are no longer copper atoms, but are something else.

Each atom has a small central nucleus consisting of sub-atomic particles called protons and neutrons. The nucleus itself is surrounded by even smaller particles called electrons, which whizz around the nucleus in what is known as an electron cloud.

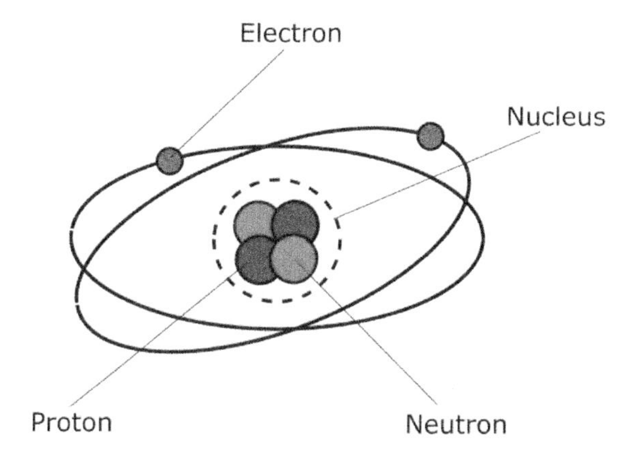

Atoms are really small, ranging in size from about 0.1 to 0.5 nanometers (1×10^{-10} m to 5×10^{-10} m)

That's roughly a million times thinner than a human hair

Protons and neutrons are even smaller, with electrons being the smallest particles

An electron is about 200,000 times smaller than a proton

Figure 1 - The structure of an atom

Electric Charge (Electromagnetism)

Protons have a positive (+) electric charge, and electrons have an equivalent negative (-) charge. Neutrons are neutral, that is, they don't have an electric charge. Although neutrons are extremely important in chemistry and physics, they are not important in electricity ... so we won't say any more about them. Electrons and protons, on the other hand, form the basis of how electric current flows ... so we will talk more about them.

Electrons are very, very small ... about 200,000 times smaller than protons. An atom of a particular element normally has the same number of protons as electrons. Copper, which is a very important element in electricity, has 29 protons in its nucleus, and 29 electrons around the outside of the atom.

Elements

There are 118 elements and each one has an atomic number, which is determined by the number of protons in its nucleus. A hydrogen atom has one proton in its nucleus, which means it has the atomic number 1. Helium has two protons in its nucleus and has the atomic number of 2. As mentioned above, copper has 29 protons in its nucleus and has the atomic number 29.

All 118 elements are listed in something called the Periodic Table which, if you've ever studied chemistry, you will know all about.

Sometimes atoms gain or lose electrons. When this happens, it causes the atom that has lost the electron to not have quite enough electrons, while the atom that has gained an electron now has slightly too many. These atoms are called ions. A positive ion occurs when an atom loses an electron (negative charge), which means that it has more protons (positive charge) than electrons. A negative ion occurs when an atom gains an extra electron so it has more electrons than protons causing it to have a negative charge.

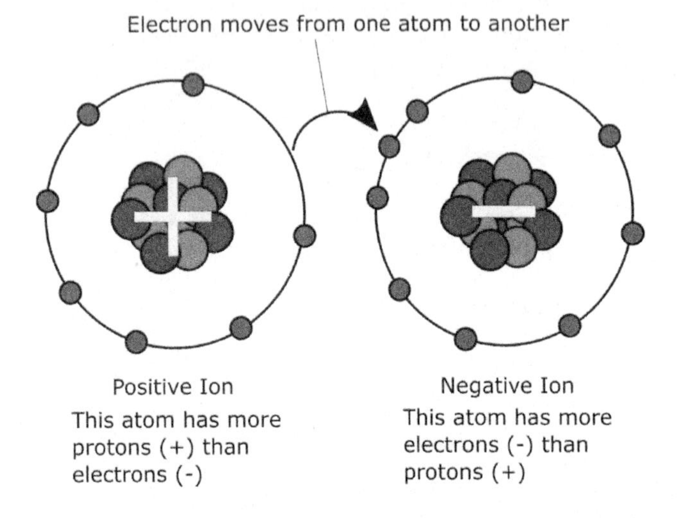

Figure 2 - Positive and negative ions

Opposites Attract

Protons and electrons have opposing electric charge, which causes them to be attracted to each other. It is this attraction that holds atoms together, with electrons being held in place in orbit around the nucleus.

The physical interaction of electrically charged particles (protons and electrons) is known as electromagnetism, which is one of the fundamental forces of nature.

This process of atoms losing and gaining electrons takes place in a random manner within an element, so some atoms are losing electrons, and some are gaining electrons, causing positive and negative ions to be created all over the place.

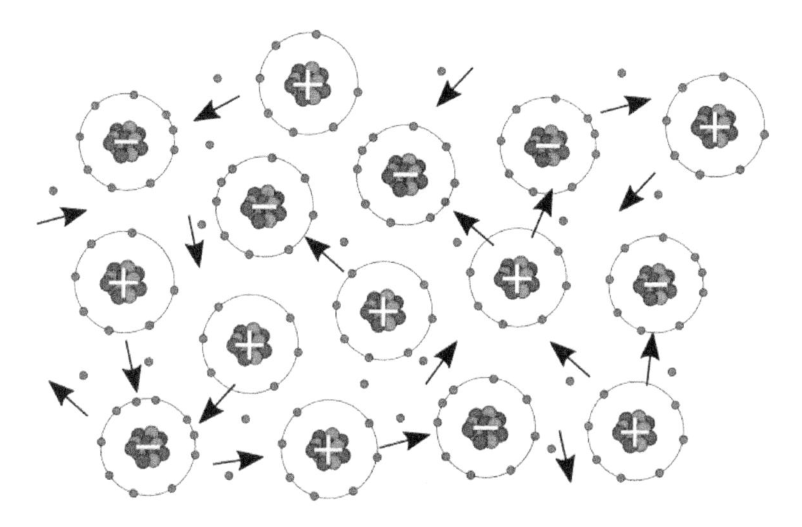

Figure 3 - The random movement of electrons between atoms

Conducting and Insulating Elements

With some elements - copper, for example - the process of electrons randomly jumping around between atoms takes place more easily than with other elements. It is this property of copper that makes it an excellent *conductor*. Other excellent conductors are silver and aluminium.

Elements that are reluctant to let their electrons move from one atom to another are known as insulators. Materials made of polymers or plastics tend to be very good insulators, which is why they are used as insulation around copper wires. Glass and paper are also excellent insulators.

No such thing as a perfect insulator

There is no such thing as a perfect insulator. This is because even insulators contain small numbers positive and negative ions, which can be used to conduct (carry) an electric current. In addition, all insulators become electrically conductive when a large enough voltage is applied such that electrons are torn away from the atoms. When this happens, it is known as the breakdown voltage of the insulator.

Voltage and Current

Although the random movement of electrons between atoms is all very interesting, it's not a lot of use on its own. However, if we could get them to travel in the same direction, then perhaps we could take advantage of that. In fact, the result of doing this is what creates an electric current within a conductor. What happens is that when all the electrons start travelling in the same direction, they begin to transfer their electromagnetic force through the conductor.

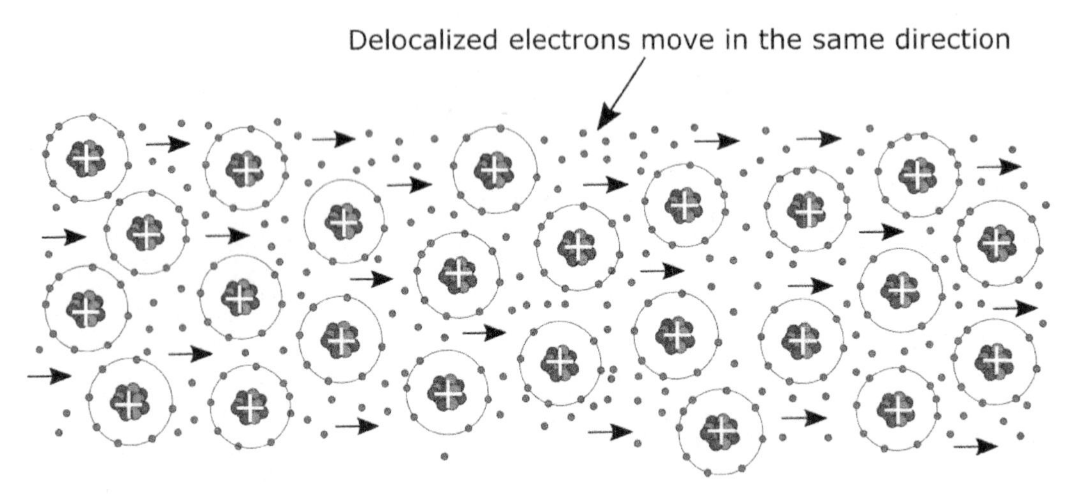

Figure 4 - Electrons all travelling in the same direction through a conductor

Although each individual electron only travels a few millimetres per second, because this is happening trillions of times simultaneously, the net result is that the flow of electrons takes place at almost the speed of light.

Getting electrons to all flow in the same direction doesn't happen by accident - we need to do something in order to make it happen. We do this by using voltage.

A voltage is the difference in electric charge between two points. So, for example, if you have one piece of metal that has an abundance of positive ions (positively charged atoms - they don't have enough electrons), and another piece of metal with an abundance of negative ions (negatively charged atoms - they have too many electrons), a voltage (difference in electric charge) exists between the two pieces. If you then attach a conductor between them, electrons (negatively charged particles) flow from one piece of metal to the other. The flow of electrons continues until all of the atoms in the circuit are electrically neutral.

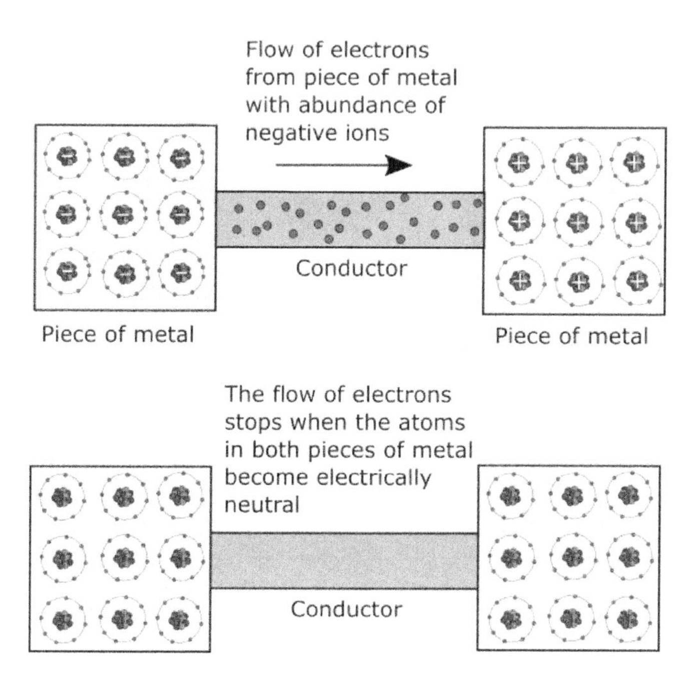

Figure 5 - Flow of electrons between two pieces of metal

Once the electrons stop flowing, they revert to their random movement between atoms as shown earlier in *Figure 3* on Page 17.

So, we can conclude that a voltage, which is a difference in electric charge between two points, needs to exist in order for electric current to flow.

Potential Difference and Electromotive Force (EMF)

Voltage can also be referred to as potential difference.

Electromotive force (EMF) is the voltage developed by any source of electrical energy, for example, a battery. It is often defined as the source's electrical potential in a circuit.

Measuring Voltage

Voltage is measured in volts (V), named after Italian physicist Alessandro Volta (1745-1827). Volta invented the voltaic pile, which was possibly the first chemical battery.

There are a few types of instrument that can be used to measure voltage:

- Voltmeter (often integrated into a multimeter)
- Potentiometer
- Oscilloscope

Voltage Sources

Batteries are often used to provide a direct current (DC) voltage for a circuit.

1.5 V Battery

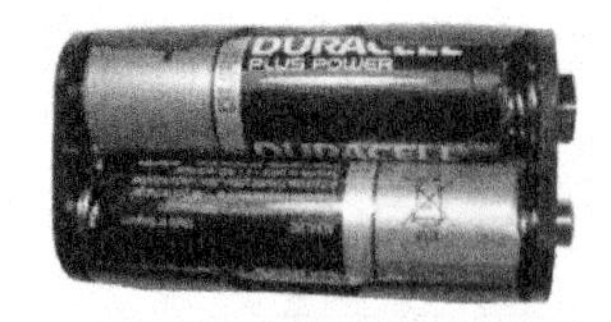

Two 1.5 V batteries connected together (in series) to provide a 3 V source

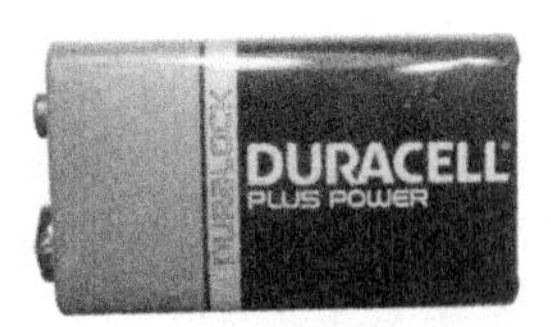

9 V Battery

Figure 6 - Selection of batteries

The voltages supplied by power companies to consumers are much higher than those provided by batteries. The exact voltage supplied to your home or business depends where in the world you live, but is either 110 to 120 volts (alternating current - AC) or 220 to 240 volts (AC).

Direct Current (DC) and Alternating Current (AC)

With direct current (DC), the movement of electric charge through a circuit is in one direction only. This is because the voltage remains constant, which keeps the electric charge flowing in the same direction. Batteries are typical sources of direct current.

In an alternating current (AC) circuit, the voltage is periodically reversed, which causes the flow of electric charge to also reverse. This normally happens about 50 or 60 times per second. AC, which is more efficient than DC at transmitting electric charge over long distances, is the form of electric power normally delivered to commercial and residential properties.

As we're dealing with electronic circuits in this book, we only look at DC, battery-driven, circuits.

Measuring Current

Current is the flow of electric charge and is measured in amperes (or amps (A)), which is a measure of how many charge carriers (electrons, in general) flow past a certain point in one second. One ampere equals a flow of 6,240,000,000,000,000,000 electrons per second.

(The ampere is named after André-Marie Ampère (20 January 1775 - 10 June 1836) who was a French physicist and mathematician.)

Ammeter

To measure how much current is flowing through a circuit you can use an ammeter, which is normally integrated into a multimeter. We discuss how to use a multimeter in Chapter 14 on Page 99.

Conventional Current Flow

As we have already established, electrons are negatively charged particles and flow when there is a difference in electric charge (voltage) between two points. The direction of flow is from the negative end of the circuit towards the positive end. This is because opposite electric charges are attracted to each other, so electrons (negative) flow towards the positive charge.

In a battery, for example, electrons flow from the negative terminal, through the circuit, and then back to the positive terminal.

However, as well as electron flow, there is something called conventional current flow, which flows in the opposite direction to electron flow.

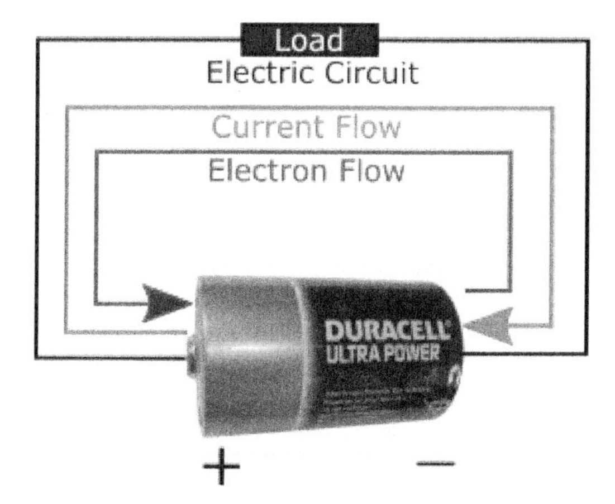

Figure 7 - Conventional current flow and electron flow

> *Note*
>
> *Conventional current flow is from positive to negative; electron current flow is from negative to positive.*
>
> *When you are trying to determine where in a circuit to place a resistor, for example, you need it to be in between the battery's positive terminal and the component that the resistor is protecting. This is because the flow of current is coming from the positive terminal.*
>
> ***Warning: Never connect the terminals of a battery directly together as this will cause the battery to overheat very quickly and may even cause it to explode. There must always be a 'load' (for example, a light bulb) in***

between the battery terminals. We discuss this later in this chapter when we look at a basic electric circuit.

Resistance

All the components contained in an electric circuit - including the wires connecting those components - resist the flow of current to a certain extent. There is no such thing as a perfect conductor so, even material like copper, offers a certain amount of resistance. Materials that are used as insulators, such as polymers, offer a lot of resistance, which is why they are used as insulators. If you mix a material that is a good conductor with a material that is a good insulator, you end up with something in between the two. This is how resistors work, which is the subject of Chapter 4 on Page 40. If you add up all the resistances of the individual components in a circuit (if they are connected in series), you end up with the overall resistance of the circuit itself.

The amount of resistance provided by a wire increases as:

- the length of the wire increases, and
- the thickness of the wire decreases.

This means that a long thin copper wire will have a tendency to provide more resistance than a short fat copper wire. However, in the experiments we carry out in this book, the jumper wires have a very little resistance and as such their resistance value can be ignored.

When electrons flow through a wire they sometimes bump into ions, which make it more difficult for the electrons to flow, causing resistance. In a long wire, this is more likely to happen than in a short wire. Likewise, in a thin wire, there are fewer electrons to carry the current than in a thick wire.

Measuring Resistance

Resistance is measured in ohms, named after the German physicist Georg Ohm who was the first person to explain the relationship between voltage, current and resistance. He did this in 1827, which was 45 years after the actual discovery had been made by Henry Cavendish. However, British scientist Cavendish failed to publish his findings, so it is ohms that are used to measure resistance.

1 ohm is the amount of resistance required to allow 1 amp of current to flow around a circuit when 1 volt is applied to that circuit.

Ohmmeter

To measure the resistance in part of a circuit you can use an ohmmeter, which is normally integrated into a multimeter. As mentioned above, we discuss how to use a multimeter in Chapter 14 on Page 99.

Ohm's Law

Ohm's Law, named after the German physicist Georg Ohm (see above) can be used to calculate voltage, current or resistance. You need to know two of these values in order to calculate the third.

Ohm's Law:

Voltage (V) = Current (I) x Resistance (Ω)

As you can see from the above formula, the Greek letter omega (Ω) is used to represent ohms. The letter I is used for current, which stands for current *intensity*. The letter V is used for voltage.

To calculate current you need to re-arrange the formula so that it becomes:

Current (I) = Voltage (V) / Resistance (Ω)

To calculate resistance you need to re-arrange the formula so that it becomes:

Resistance (Ω) = Voltage (V) / Current (I)

Why is the symbol for current 'I' and not 'A'?

The symbol for current is I, which may seem a bit strange. It originates from the French phrase intensité de courant, or in English, current intensity.

Power

Power is the amount of work done by an electric circuit and is measured in watts (W).

A battery can be used as a power source for a small electric/electronic circuit. However, to supply power to homes and businesses, electric generators are used. Electrical power can be carried over long distances and then efficiently converted into other forms of energy such as light, heat or motion.

The amount of power generated by an electric circuit depends on the amount of current passing through the circuit, as well as the amount of voltage used to push the current through the circuit. This can be represented using the following formula:

Power (W) = Voltage (V) x Current (I)

So, for example, a current of 1 A, which is being pushed around a circuit by a 1 V power supply, generates 1 W of power.

In order to do something useful, the electric energy carried by the current needs to be converted into a different form of energy such as heat or light. For example, with an electric heater, a voltage pushes current through heating elements to generate heat; with a light bulb, a voltage pushes current through a filament to generate heat and light. When this happens, the circuit is said to *dissipate* a certain amount of power in watts. For this reason, heaters and lights are both given power ratings in watts, for example, a 2 kilo-watt (kW) heater, a 40 W light bulb, and so on. The higher the power rating, the more heat or light is generated by the circuit, in other words, the more work is being done.

In the same way that you can calculate the power generated by a circuit if you know the voltage and current, you can calculate the current flowing through a circuit if you know the power and voltage. To do this you just need to re-arrange the above formula:

Current = Power / Voltage

Or, to calculate voltage when you know the power and current:

Voltage = Power / Current

A Basic Electric Circuit

An electric circuit is a complete loop that starts and ends at the same point and provides a conductive path for an electric current to flow.

Figure 8 (below) shows a very basic circuit in which a battery is connected to a light bulb.

Figure 8 - A very simple electric circuit

In this circuit, current flows from the battery to the light bulb. The current then continues through the bulb and returns to the battery, completing the circuit.

Electrical connectors on a light bulb

A light bulb has two electrical connectors (terminals) on it: one is at the end of the base and is normally a round dot, and the other is on the side of the base, which is sometimes a thread. It doesn't matter whether you connect the battery's positive terminal to the side of the light bulb or the bottom - the circuit will work either way.

Things to note about the circuit:

- **Voltage** - The battery supplies a voltage that causes current (electrons) to flow around the circuit. The electrons leave the negative terminal of the battery and travel around the circuit (through the light bulb) until they reach the positive terminal of the battery.
- **Closed Circuit** - The circuit is closed - it forms a complete loop, starting and ending at the same place (the battery). If there was a break (or gap) in the circuit it would not work.
- **Load** - In order to do something useful, an electric circuit needs to have a load. In our case, the load is the light bulb. In electronic circuits, the load is made up of components such as resistors, capacitors, light emitting diodes (LEDs), buzzers, motors, and light sensors.

Adding a Switch

In its current form, the circuit is not particularly practical because there is no way of turning off the light bulb without disconnecting the battery. To address this short-coming we can add a switch, as shown in *Figure 9* (below).

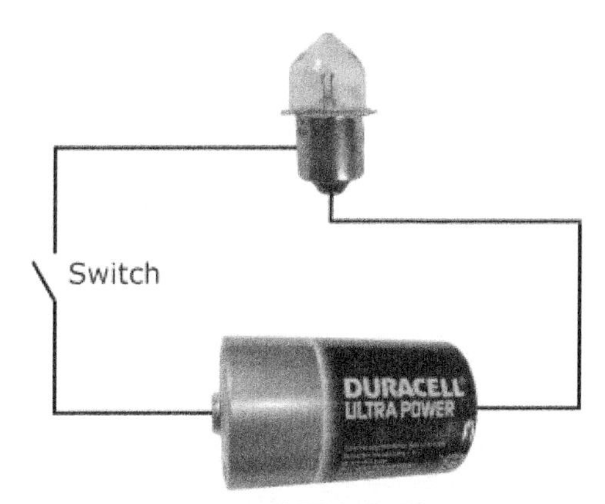

Figure 9 - Adding a switch to the circuit

What we have done by introducing the switch is to create a situation whereby we can turn the closed circuit into an open circuit. This prevents the current from flowing through the light bulb, which turns it off.

Short Circuit

Electric circuits can become very complicated and ensuring that the current flows around those circuits correctly is a tricky business and sometimes *short circuits* occur. This is when the current takes an alternative route to the one you want it to, and in doing so incorrectly bypasses some or all of the components that make up the system. The circuit we have been looking at so far is very simple, and introducing a short circuit is very easy to do, as shown below.

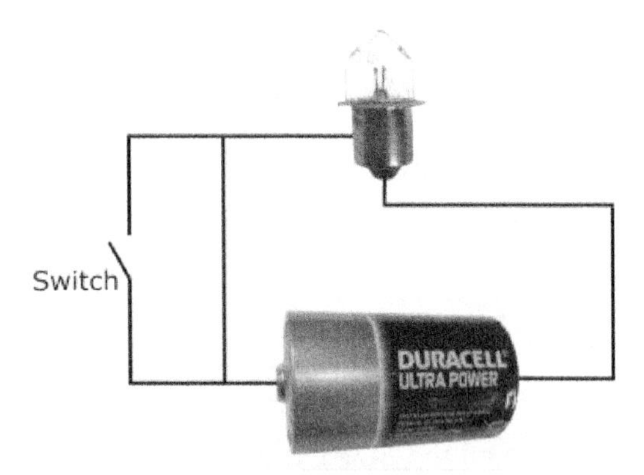

Figure 10 - A short circuit

The short circuit shown in *Figure 10* (above) causes the switch to be bypassed, meaning that whether it is open or closed, the light bulb will always be on. Although this short circuit would be annoying if you had designed the circuit, it would not cause any damage to the light bulb as we have effectively re-created the original circuit shown in *Figure 8* on Page 24.

When the switch is closed, the current will flow down both paths, which then converge on their way to the light bulb.

Now consider the circuit shown below in *Figure 11*.

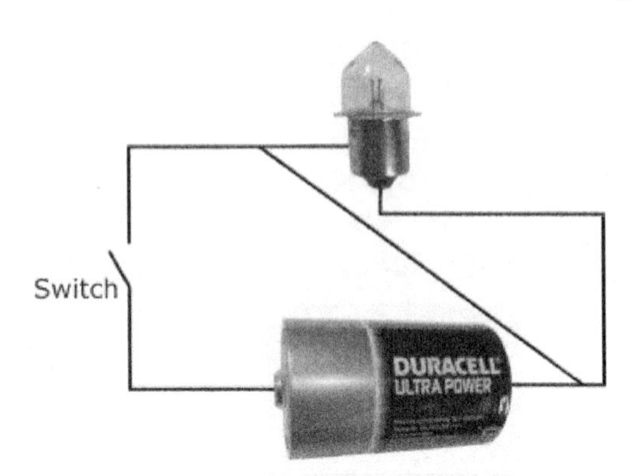

Figure 11 - A more serious short circuit

The short circuit shown above is more serious because it bypasses the light bulb (the circuit's load). When the switch is closed, the circuit will get very hot as the current races between the two terminals of the battery. Some current will flow through the light bulb when the switch is closed, but most of it won't.

Why do things get hot?

A battery has a very small amount of internal resistance. We can use Ohm's law to calculate how much current will flow around the circuit when the two battery terminals are connected to each other. (I'm assuming that the resistance of the wire is negligible so I haven't included it in the calculation.)

volts = current x resistance

Therefore:

current = volts / resistance

volts = 1.5 V

resistance = 140 mΩ (this is the approximate internal resistance of the battery)

Therefore:

current = 1.5 V / 0.140 Ω

Current = 10.71 amps

This is a lot of current to pass through the battery, which will cause a very rapid build up of heat in the circuit and may even cause the battery to explode.

The Battery

As with virtually all of the circuits we describe in this book, the voltage source (the power source) is provided by a battery. The battery shown in the above circuits is a 1.5 volt 'D'. If we add a second battery in series with the first one, the light bulb will glow more brightly because we have increased the amount of voltage pushing the current around the circuit.

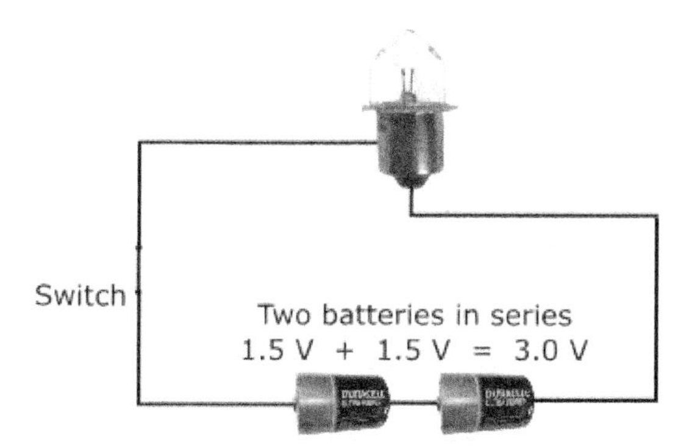

Figure 12 - Two Batteries in Series

On the other hand, if we add more light bulbs to the circuit, they will become dimmer because the voltage is having to push the current through a bigger load.

Electronic Circuits

Electronic circuits are made up of various components that work together to create a circuit that does something. Exactly what that "something" is depends on the circuit. Electronic components can be categorized into two groups:

- Discrete components
- Integrated circuits (ICs)

Discrete components include items such as resistors, capacitors, and diodes. Integrated circuits (ICs) are tiny circuits etched onto silicon and are often referred to as chips. They are normally used to perform a particular function, for example a timer, and can be combined with discrete components to form larger circuits.

ICs have pins on them that provide access to the various parts of the IC's internal circuit. Getting the IC to function properly requires correct use of these pins.

Schematic Circuit Diagrams

Although the circuit diagrams we have looked at so far in this chapter look quite nice in that you can easily recognize the battery and the light bulb, once a circuit starts to become more complicated, this approach would soon become quite cumbersome. So, instead, schematic circuit diagrams are normally used in which components are represented by symbols, with lines between them to show how they are connected.

A schematic diagram for the lighting circuit is shown below in *Figure 13*.

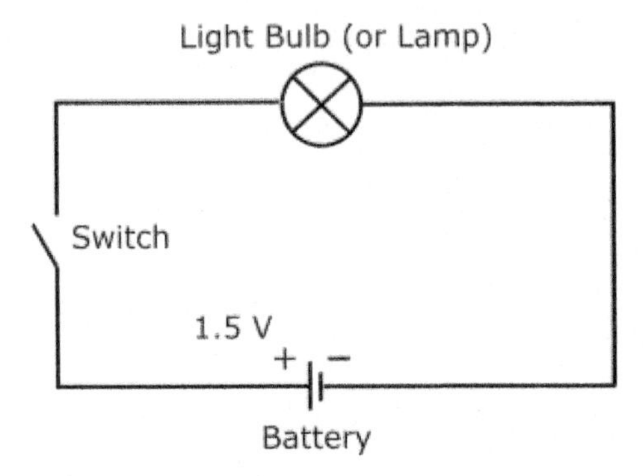

Figure 13 - A simple schematic circuit diagram

The idea is that the symbols convey enough information to enable someone to understand and build the circuit, but do not contain any unnecessary detail regarding the components themselves.

In *Figure 13* on the previous page, we have three symbols:

- The battery (1.5 V),
- the light bulb (or lamp), and
- the switch.

The lines that join the symbols represent wires on a breadboard or traces of copper on a printed circuit board.

A couple of things to note about schematic circuit diagrams:

- They always depict conventional current flow (positive to negative) rather than electron current flow (negative to positive).
- The layout of components in a schematic circuit diagram does not have to match the physical layout of the components when the circuit is built.

Chapter 2

Equipment You Need in Your Electronics Laboratory

To carry out electronics experiments you need to have a variety of electronic components and equipment available to you. The more components you have in your set, the more you will be able to play around with circuits, but initially you'll be able to get by with a fairly modest set.

The list of components and equipment in the following table is reasonably extensive, so don't worry about getting everything straight away. In fact, it's probably better to acquire stuff slowly rather than going out and buying everything in one go. That way, you'll be able to spend more time getting components and equipment that are right for you.

Components and Equipment	Description
Breadboard(s)	A breadboard enables you to prototype electronic circuits without the need to solder components in place. With a breadboard you can insert a component into position and then remove it and put it somewhere else. Beneath the holes are a series of internally connected rows and columns that enable components to be connected to each other by jumper wires. To supply power to the breadboard you normally use a battery. To carry out the experiments in this book you need to have at least one small breadboard like the one shown here. I find it useful to have a few breadboards though so that I don't have to take the breadboard apart every time I want to build a new circuit.
Battery (or Batteries)	Almost all of the electronic circuits described in this book use a battery (or batteries) as the power source. When I'm working on circuits myself, I generally have two

 9 V Battery 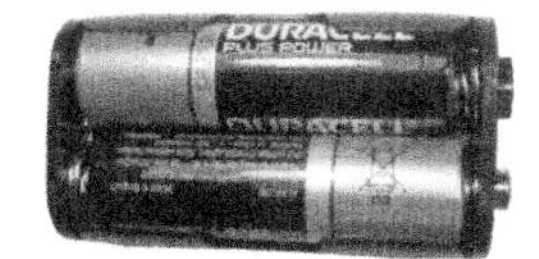 Two 1.5 V AA Batteries in a Battery Box 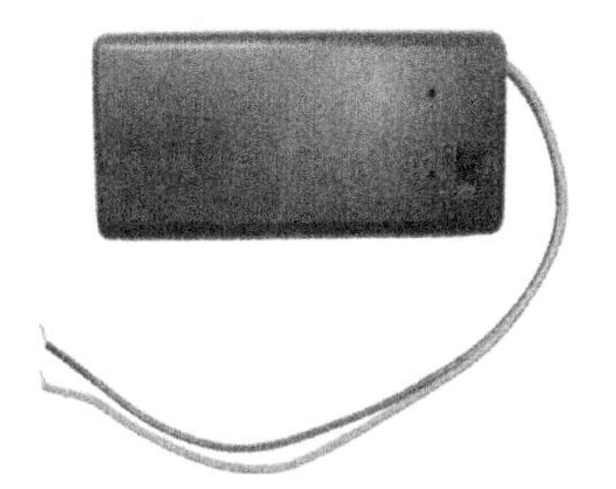 9 V Battery Box (with On/Off Switch) 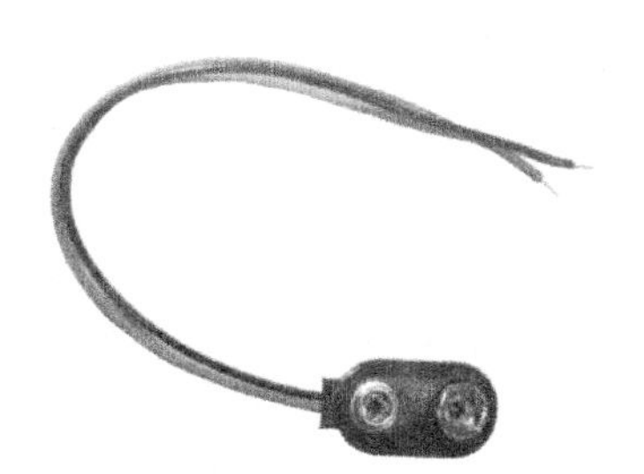 Battery Clip	battery configurations available: • a single 9 V battery in a box (with an On/Off switch), and • two 1.5 V AA batteries connected in series with each other (to produce an output of 3 V), contained in a non-switchable box. With this battery configuration I normally use a battery clip to connect the battery to the breadboard (see image to the left). Most of the time I use the 9 V battery in the switched box because it enables me to easily switch the power to the circuit on and off while I'm testing things, rather than having to physically disconnect one of the power supply wires from the breadboard. I find it useful to have a 3 V power supply as well though so that I can try circuits out at this lower power rating. You will learn later in the book that using a 9 V supply when a much smaller supply voltage will suffice means that power is lost (wasted) through heat dissipation. Another option is to use rechargeable batteries, which are often 1.2 V. You would normally connect three or four of these in series (in a battery box) in order to achieve a sufficient voltage supply.
Resistors 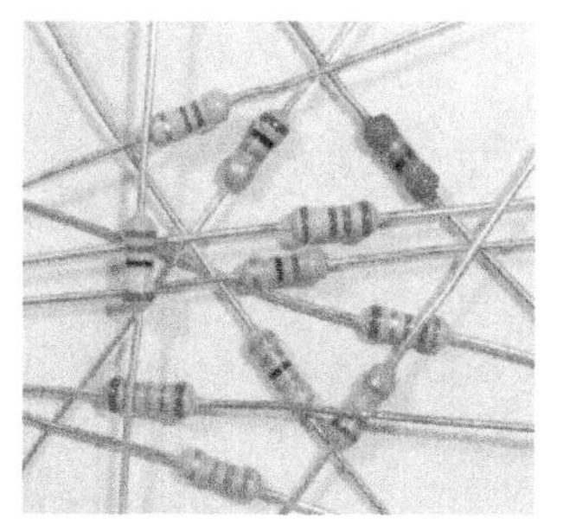 Selection of Resistors	Pretty much every electronics circuit uses resistors, which are small devices used to control the flow of current in a circuit. An electronic circuit is made up of components that require a certain amount of current in order to operate correctly. Resistors enable you to control how much current flows into components to ensure that they work as intended and do not get damaged. Resistors are very cheap and when you start building circuits you will notice that you very quickly accumulate a large selection of resistors. In my opinion though you should avoid buying resistor "lucky

<table>
<tr>
<td>
Variable Resistor
(Potentiometer)

Light Dependent Resistor (LDR)
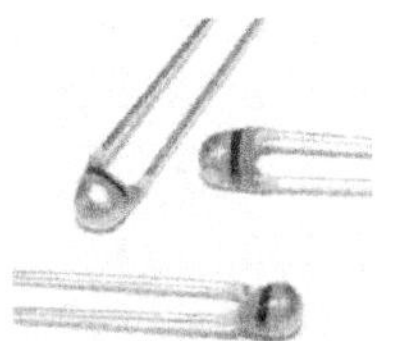
Thermistors</td>
<td>bags" as you may well end up with a lot of resistors that you never use.

Instead, buy a range of resistors, for example, ten 47 Ω resistors, ten 470 Ω resistors, ten 1 kΩ (1000 Ω) resistors, and ten 2.2 kΩ resistors (don't worry, we explain all about resistance values when we cover resistors later in the book).

You should also get yourself a few variable resistors (potentiometers). With this type of resistor you can manually adjust the resistance value from a minimum to a maximum value, for example from just above 0 Ω to 50 kΩ.

We also use a Light Dependent Resistor (LDR) when we design a light sensor (night light), so you should get at least one of these. With an LDR, the amount of resistance provided by the resistor varies with the amount of light detected.

You could also consider getting a few thermistors. A thermistor is a resistor in which the amount of resistance provided depends on the temperature.

Resistors of all types are cheap, so you may as well get yourself a good selection to use in your circuits.</td>
</tr>
<tr>
<td>Diodes
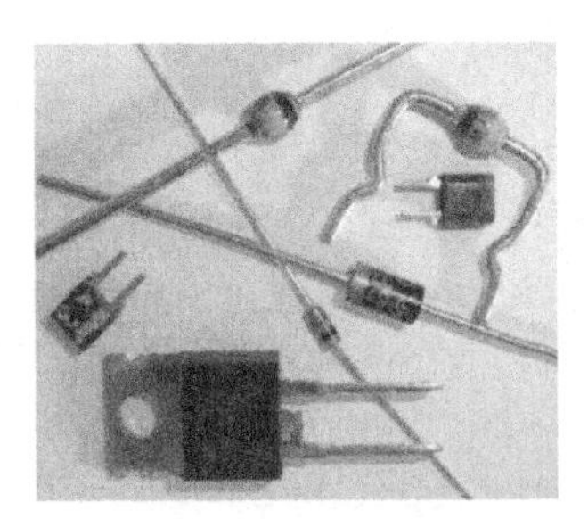</td>
<td>A diode is an electronic component that lets current flow in one direction only.

A diode has two leads (or pins): the anode and the cathode. You can normally identify which is which because there will be some kind of marker on the diode such as a silver or coloured band, which is next to the cathode.

If the anode is connected to a higher voltage than the cathode, current flows through the component (from the anode to the cathode). If the cathode is connected to a higher voltage than the anode, current does not flow through the component.

Diodes are very cheap to buy so you should get yourself a selection of them for your electronics laboratory.</td>
</tr>
</table>

Light Emitting Diodes (LEDs) 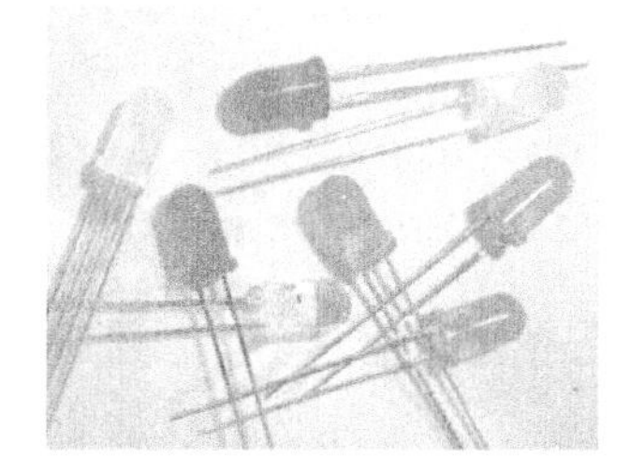	As with normal diodes (mentioned above), light emitting diodes (LEDs) have an anode and cathode and only allow current to flow in one direction through the component. However, different to normal diodes, LEDs emit light when current passes through them. LEDs, which come in a variety of colours, are very cheap so you may as well get a selection of them. Some of them even glow in multiple colours. Most of the experiments in this book involve an LED of some sort.
Capacitors 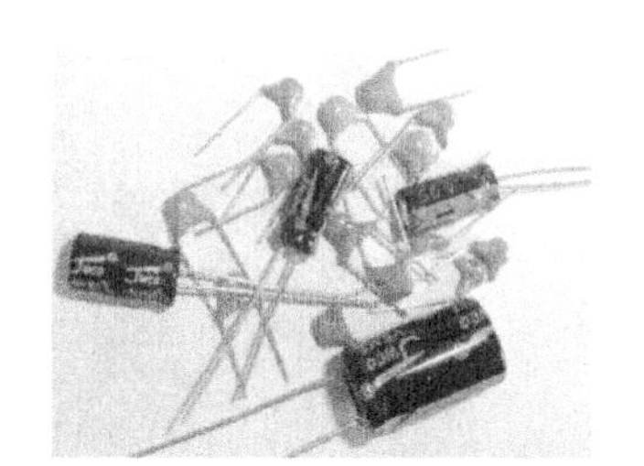	Capacitors are used to store electrical energy and are very widely used in electrical and electronic systems. Capacitors come in different sizes, with large ones able to store more electrical energy than small ones. They can also be polarized (generally the larger ones) or non-polarized (generally the smaller ones). A polarized capacitor has a positive pole and a negative pole and has to be inserted the correct way round in a circuit in order to work. As with the other components we have mentioned, capacitors are not expensive at all so get yourself a good selection.
Transistors  Selection of Transistors	Although all electronics components are important, there is something special about the transistor because it revolutionized the electronics industry when it was invented by John Bardeen, Walter Brattain and William Shockley in 1947. Transistors are the building block of modern electronic devices and are used to switch or amplify electronic signals. Although all of the transistors we use in this book are big enough to be able to physically pick up and handle, they can be extremely tiny, with modern micro chips containing millions of them. Again, they're cheap, so buy a selection of them.

Push Buttons / Switches 	A push button is a switch mechanism that lets you open or close a circuit. In Chapter 20 on Page 130 we design and build a circuit that charges and discharges a capacitor. In order to do this we use two push buttons: one that controls the *charging* part of the circuit, and one that controls the *discharging* part.
Jump Wires 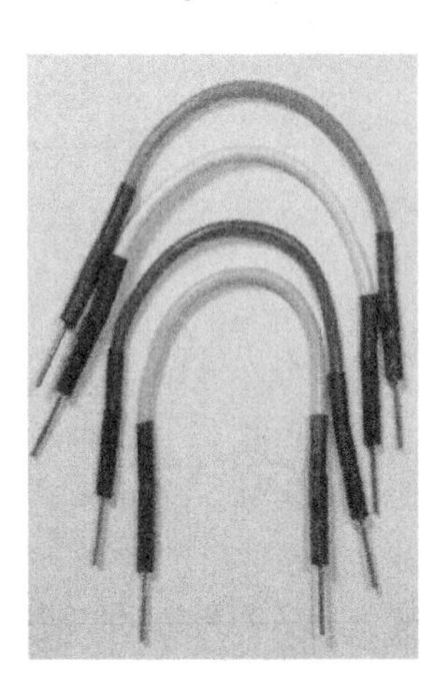	A jump (or jumper) wire is an electrical wire with a connector at each end. Sometimes the wires can be grouped together within a cable, but in this book we will just use individual jump wires. The connectors can be inserted into the holes on a breadboard to connect components and create a circuit. The thickness of the wires is normally 22 AWG (0.33 mm^2). AWG stands for American Wire Gauge, which is a standardized wire gauge system that has been in use since 1857. Different coloured wires are available. The colours can be used to indicate different signals or paths if required, but as the wires are the same thickness, they are all the same from an electrical point of view. Different lengths of wire are also available. In order to carry out the experiments in this book you need to have a few jump wires.
Crocodile Clips 	A crocodile (or alligator) clip is a sprung metal clip with long, serrated jaws which is used for creating a temporary electrical connection. It gets its name from the resemblance of its jaws to that of a crocodile's. Crocodile clips are very useful when you're using a multimeter to measure current, voltage, or resistance in a circuit because sometimes you feel as though you need an extra pair of hands when you're trying to take measurements.

Small DC Motor  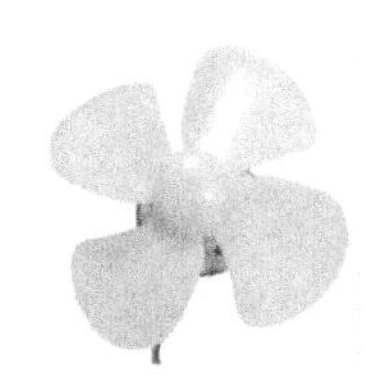	Direct Current (DC) motors are good fun to play with so you should certainly get at least one as part of your electronics laboratory. Ideally, look for something that operates in the 6 to 15 V range so it is suitable for use with a 9 V battery. You should also get a fan blade that you can attach to the motor's drive shaft. This not only enables you to see when the motor is running, but it also lets you generate electricity using wind power, which is subject of Chapter 18 on Page 122.
Terminal Block 	A screw terminal block is a type of electrical connector where a wire is held by the tightening of a screw. We use one in this book when we're building circuits that use a DC motor in Chapter 17 on Page 116 and Chapter 18 on Page 122.
Multimeter 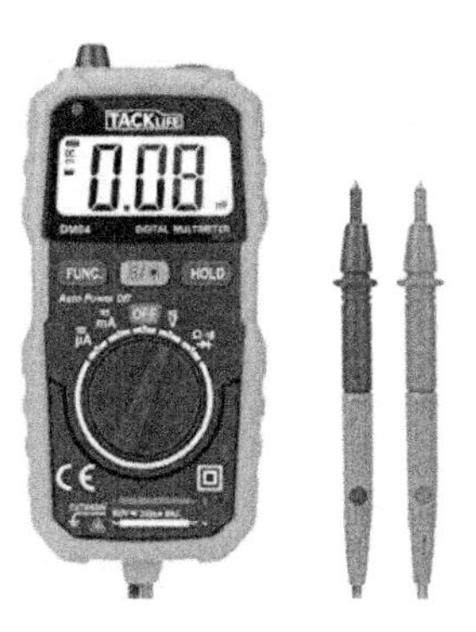 Digital Multimeter  Analog Multimeter	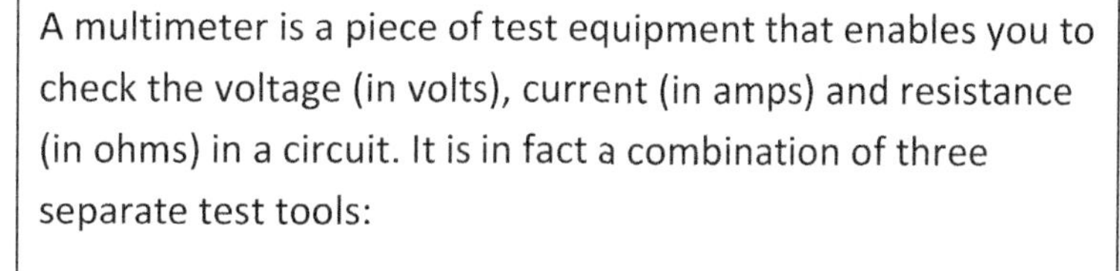 A multimeter is a piece of test equipment that enables you to check the voltage (in volts), current (in amps) and resistance (in ohms) in a circuit. It is in fact a combination of three separate test tools: • an ammeter, which measures current, • a voltmeter, which measures voltage, and • an ohmmeter, which measure resistance. There are two types of multimeter: • digital, which has an LCD screen that gives a decimal read out, and • analog, which has a display whereby a bar moves across a scale of numbers, and must be interpreted. One other thing to consider is that digital multimeters can be auto ranging. What this means is that once you set the multimeter to read current, for example, it doesn't matter whether you are measuring a few milliamps or a few amps - the device automatically adjusts itself to cope with this. With a multimeter that is not auto ranging you need to select an

	appropriate range before taking a measurement. The type of multimeter you decide to use is up to you. Prices range quite considerably, but you can pick up a small auto ranging multimeter for not very much money.
Oscilloscope	An oscilloscope is a type of electronic test instrument that allows observation of varying signal voltages, usually as a two-dimensional plot of one or more signals as a function of time. Oscilloscopes can be very expensive and I would not recommend splashing out on one – certainly not an expensive one – until you're sure you want to take your electronics hobby seriously. However, you can pick up a small oscilloscope such as the DS202 (shown here) for around £100 ($100 - $130).
Magnifying Glass	Reading the markings on small electronics components can be a bit tricky at times, especially when you're as old as I am with failing eyesight! A magnifying glass makes life a lot easier. Getting something with a magnification of about 6X should suffice, and you shouldn't need to spend very much.
Solar Panel	Playing around with a solar panel is good fun, and a small one with specs something in the region of O.5 W, 5 V, 100 mA, is also very cheap to purchase. We'll examine how to measure the current and voltage generated by a solar panel, and how to connect them together to increase either the current or the voltage output.
Storage Unit	A storage unit for your components and equipment is essential. Look to spend in the region of £20 ($20 - $25) or perhaps a bit more to get something that will accommodate all your equipment and components. Label the draws carefully, and sub-divide them into smaller

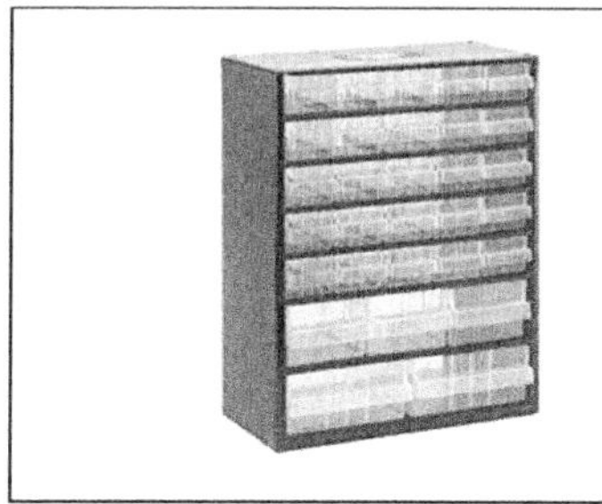	compartments if you can (a lot of storage units come with small plastic inserts that enable you to do this).

Table 1 – Stocking your electronics laboratory

Chapter 3

Avoiding Injury

Working with electricity brings with it two big dangers: the risk of being electrocuted and the risk of starting a fire. As we will only be using batteries of no more than 9 V in all of our experiments, you are unlikely to injure or kill yourself as a result of an electric shock. Burning yourself or even causing a fire are a different matter though. There is always the risk of causing a short circuit between the two poles of a battery when you are creating electric circuits, which can very quickly generate a lot of heat as the current flows uncontrollably between the two poles. A short circuit will cause the wires, components and the battery itself to become very hot and perhaps even explode.

General Guidelines

Here are a few general guidelines to help you to stay safe:

- Always connect up all the components in your circuit before you supply any power to it.
- Once power is being supplied, touch the battery and other components to check that nothing is getting hot. Do this on a regular basis while you are working on the circuit.
- When you have finished working on the circuit, disconnect the battery from the circuit.
- If you have battery clip connected to the battery terminals, remove it when the battery is not supplying power to a circuit. If the clip's two wires touch each other while the clip is connected to the battery, this will cause a short circuit and the battery will get very hot, very quickly.
- Disconnect the power source straight away if you smell burning.
- Wear protective glasses just in case things go wrong and your battery explodes.
- Try to avoid working in your electronics lab when there is no-one else in the house.

Hidden High Voltages

Depending on the circuits you build you may well create situations whereby hidden high voltages exist in the circuit. Pay special attention to any circuits that contain capacitors - especially big ones - as these can store electrical energy long after the battery power supply has been removed from the circuit.

Things to Have to Hand

Here are a few things to have to hand when you're building circuits:

- Fire extinguisher
- First aid kit
- Cell (mobile) phone in case you need to call for help

Chapter 4

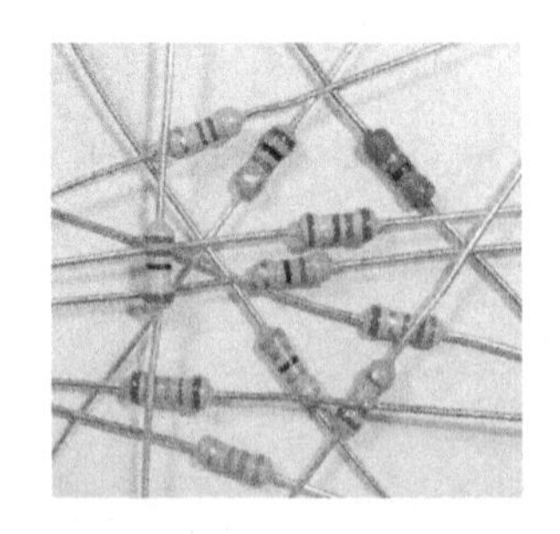

Resistors

Resistors are small devices used to control the flow of current in a circuit, and are used in pretty much every electronics circuit. They are made from a combination of material that conducts current easily (for example, a conductor such as carbon) and material that does not conduct current easily (for example, an insulator such as ceramic). The mixture of these two materials determines the amount of resistance provided by the resistor. A mixture high in ceramic produces a resistor that imposes a high resistance on the flow of current, while a mixture high in carbon produces a resistor that lets current flow relatively easily.

Why Resistors are Important

An electronic circuit is made up of components that require a certain amount of current in order to operate correctly. If the amount of current is too low, the component may not work properly or perhaps not at all. For example, if a light emitting diode (LED) requires 25 milliamps of current in order to illuminate properly, but the circuit is only supplying 10 milliamps, the LED will not be very bright. On the other hand, if you allow too much current to flow through a component, you run the risk of destroying that component. For an LED it might mean that it shines very brightly and doesn't last very long, or it might even cause the LED to go bang.

By using Ohm's Law you can calculate the size of resistor you require in order to allow the correct amount of current to flow through a component.

Resistor Sizes

In 1952 the IEC (International Electrotechnical Commission) defined a standard for resistance values and tolerances to ease the mass production of resistors. These are referred to as preferred values or E-series, and are published in the IEC 60063:1963 standard.

The following E series exist:

- E6 series (tolerance 20%)
- E12 series (tolerance 10%)
- E24 series (tolerance 5% and 1%)
- E48 series (tolerance 2%)
- E96 series (tolerance 1%)
- E192 series (tolerance 0.5%, 0.25% and 0.1%)

Don't worry about what resistor tolerance means at this stage, we cover that later on.

The following table shows the resistor values that conform to the E12 series, which is the most popular of the E series.

1.0	10	100	1.0 k	10 k	100 k	1.0 M
1.2	12	120	1.2 k	12 k	120 k	1.2 M
1.5	15	150	1.5 k	15 k	150 k	1.5 M
1.8	18	180	1.8 k	18 k	180 k	1.8 M
2.2	22	220	2.2 k	22 k	220 k	2.2 M
2.7	27	270	2.7 k	27 k	270 k	2.7 M
3.3	33	330	3.3 k	33 k	330 k	3.3 M
3.9	39	390	3.9 k	39 k	390 k	3.9 M
4.7	47	470	4.7 k	47 k	470 k	4.7 M
5.6	56	560	5.6 k	56 k	560 k	5.6 M
6.8	68	680	6.8 k	68 k	680 k	6.8 M
8.2	82	820	8.2 k	82 k	820 k	8.2 M

Table 2 - E12 Resistor values

In the above table 'k' stands for kilo-ohm (1,000 ohms), and 'M' stands for mega-ohm (1,000,000 ohms).

To see a list of resistors values that conform to the other E series, take a look at this web page:

http://www.resistorguide.com/resistor-values/

Resistor Colour Coding

A resistor has a series of coloured bands around it that indicate the amount of resistance provided by the resistor (the value of the resistor), as well as its tolerance, which is a measure of its accuracy - in other words, how close to the indicated resistance value the resistor actually is.

Figure 14 - A 470 kΩ (470,000 Ω) 10% Tolerance E12 Standard Resistor

Looking at the resistor in *Figure 14* we can see that it has four bands:

- Yellow
- Violet
- Yellow
- Silver

The majority of resistors have four bands; the first three indicate the resistance value and the fourth indicates the tolerance. If a resistor has five bands, the first four represent the value and the fifth indicates the tolerance.

It can sometimes be tricky to determine from which side you should read the colours. The way to do it is to try to determine the band that is closest to the end of the resistor and start with that colour. If the resistor has a gold or silver band on it (as is the case for the resistor shown above in *Figure 14*) then you should position that band on the right-hand side when you are working out the resistor's value. This is because resistor values never start with a gold or silver band, so those colours can never be on the left-hand side.

There are plenty of online calculators to help you work out resistor values and colours - whether you have a resistor and you're not sure what its value is, or if you know what value you want but you're not sure what the resistor looks like. You can find a really good calculator here:

http://www.hobby-hour.com/electronics/resistorcalculator.php

Tip

You can also use a multimeter to quickly measure the resistance of a resistor. Refer to Chapter 14 on Page 99 for more information about how to do this.

Colour	Digit 1	Digit 2	Digit 3*	Multiplier	Tolerance
Black	0	0	0	x 10^0	
Brown	1	1	1	x 10^1	± 1% (F)
Red	2	2	2	x 10^2	± 2% (G)
Orange	3	3	3	x 10^3	
Yellow	4	4	4	x 10^4	
Green	5	5	5	x 10^5	± 0.5% (D)

Blue	6	6	6	x 10^6	± 0.25% (C)
Violet	7	7	7	x 10^7	± 0.1% (B)
Grey	8	8	8	x 10^8	± 0.05% (A)
White	9	9	9	x 10^9	
Gold				x 0.1	± 5% (J)
Silver				x 0.01	± 10% (K)
None					± 20% (M)

Table 3 - Standard Resistor Colour Code Table

** This digit only applies to 5-band resistors.*

Table 4, below, shows a few example resistance values.

Resistor	Value	Tolerance	Band Colours	N° of Bands	Calculation
	470 Ω	5%	Yellow, Violet, Brown, Gold	4	1st Digit = Yellow = 4 2nd Digit = Violet = 7 Brown = x 10^1 Gold = ± 5% Therefore, the value is: 47 x 10 = 470 Ω (± 5%)
	10 kΩ	2%	Brown, Black, Black, Red, Red	5	1st Digit = Brown = 1 2nd Digit = Black = 0 3rd Digit = Black = 0 Red = x 10^2 Red = ± 2% Therefore, the value is: 100 x 10 x 10 = 10000 Ω (± 2%)
	1 kΩ	20%	Brown, Black, Red	3	1st Digit = Brown = 1 2nd Digit = Black = 0 Red = x 10^2 No colour = ± 20% Therefore, the value is:

					10 x 10 x 10 = 1000 Ω (± 20%)
	2.2 MΩ	10%	Red, Red, Green, Silver	4	1st Digit = Red = 2 2nd Digit = Red = 2 Green = x 10^5 Silver = ± 10% Therefore, the value is: 22 x 10 x 10 x 10 x 10 x 10 = 2200000 Ω (± 10%)

Table 4 – Example resistance values

Resistor Tolerance

We have established from the previous sections that as well as having a resistance value in ohms, resistors also have a tolerance rating. What this means is if you use a 100 Ω resistor that has a 10% tolerance, the actual resistance value will be somewhere between 90 Ω and 110 Ω. In many electronics circuits a tolerance of 5% or 10% is acceptable, and that is certainly the case with all the circuits we use in this book. If, when you design and build your own circuits, you need a higher degree of accuracy (in other words, a smaller tolerance), you can simply use an E series resistor that matches your requirements. Here is the list of E series resistors (repeated from earlier):

- E6 series (tolerance 20%)
- E12 series (tolerance 10%)
- E24 series (tolerance 5% and 1%)
- E48 series (tolerance 2%)
- E96 series (tolerance 1%)
- E192 series (tolerance 0.5%, 0.25% and 0.1%)

Note that you can get resistors outside of these standard series, for example, you can buy E12 series resistors that have a tolerance of 5%.

Why are there different tolerances?

Manufacturing resistors with a small tolerance is more expensive than manufacturing them with a large tolerance. Resistors with a tolerance of 10% are cheaper to make - and therefore, buy - than resistors with a tolerance of 0.1%.

Resistor Power Rating

One more thing to consider when using resistors is their power rating, which is measured in watts (W) and will typically be 1/8 (or 0.125) W or 1/4 (or 0.25) W.

The way a resistor works is to restrict the flow of current that passes through it, which causes the resistor to heat up. If the resistor heats up too much it will burn up. The amount of power a resistor can handle before it burns up is calculated using the following formula:

power = voltage x current

As an example, let's assume that a 470 Ω is being used in a circuit, and it has 9 V across it. Using Ohm's Law we can calculate how much current will flow through the resistor:

voltage = current x resistance

Therefore:

current = voltage / resistance

current = 9 / 470

current = 0.019 amps (or 19 milli-amps)

Now that we know the voltage (9 V) and current (0.019 A), we can calculate the power that will be dissipated by the resistor:

power = 9 x 0.019

power = 0.172 watts

For a resistor with a power rating of 1/8 (0.125) W, this would be too much power to dissipate. However, for a 1/4 (0.25) W resistor, this would be acceptable.

Identifying the Power Rating

When you buy a pack of resistors, the power rating will be stated on the packaging. However, if you have a box of assorted resistors just lying around, you can't tell what the power rating of a resistor is by looking at it - unlike the resistance value and tolerance rating, there is no indication on the resistor as to its power rating. You can, however, get an idea of a resistor's power rating based on how big the resistor is - bigger resistors have bigger power ratings.

Potentiometers

All of the resistors we have looked at so far have had a fixed resistance, for example, 470 Ω, 2.2 kΩ, and so on. It is possible though to have a resistor that has a variable resistance; such resistors are called potentiometers (or *pots*, for short). Typical uses of a potentiometer are:

- volume control on a radio
- dimmer switch for a light
- thermostat

> *Potentiometers, Rheostats and Variable Resistors*
>
> *A potentiometer can be used in two ways in a circuit:*
>
> - *as a variable resistor (rheostat), or*
> - *as a voltage divider.*
>
> *As this chapter of the manual deals with resistors, we will just concentrate on how potentiometers are used as variable resistors.*

Using a Potentiometer as a Variable Resistor (Rheostat)

To use the potentiometer as a variable resistor, only two pins are used: one of the outside ones and the center pin. The position of the wiper determines how much resistance the variable resistor provides.

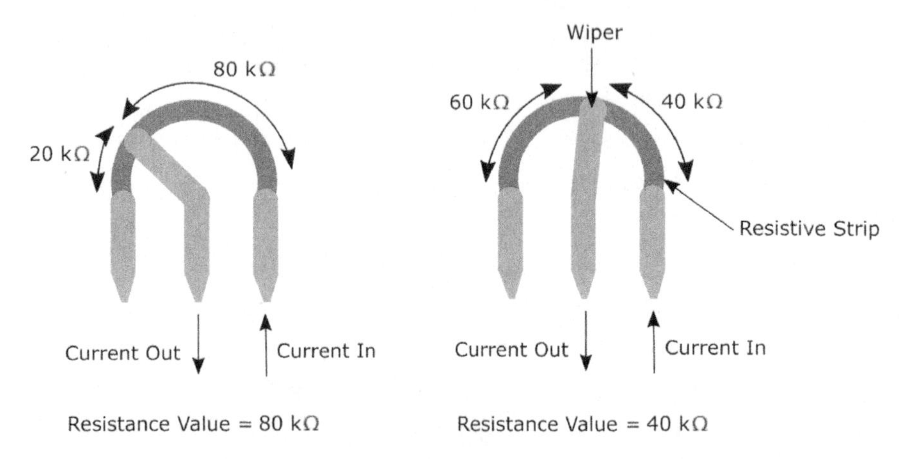

Figure 15 - A 100 k Ω variable resistor

A 100 kΩ variable resistor, as shown in *Figure 15*, has a maximum resistance of 100 kΩ and a minimum of 0 Ω. This means that by changing the position of the wiper, the resistance provided is somewhere between 0 Ω and 100 kΩ.

To adjust the resistance of a variable resistor, you normally turn (rotate) a knob, but with some variable resistors you need to move a slider.

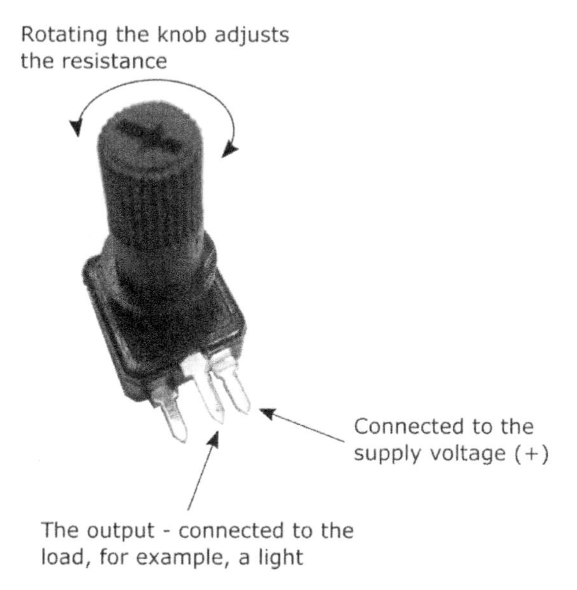

Figure 16 – A potentiometer used as a variable resistor

Variable Resistor Ratings

As with normal 'fixed-value' resistors, a variable resistor has a resistance value, for example, 100 kΩ. The way this works is that the resistance value between one of the fixed terminals and the wiper terminal, plus the resistance value between the other fixed terminal and the wiper terminal, always add up to the total resistance value of the resistor.

Figure 15 on the previous page shows the wiper in two different positions on the resistive strip. In both cases, the values either side of the wiper terminal add up to 100 kΩ.

Linear Versus Logarithmic Scales

Figure 15 assumes that the scale between 0 and 100 kΩ is a linear one. In other words, if the wiper is 25% of the way round the resistive strip, the resistance value to the left of the wiper terminal is 25 kΩ and the value to the right is 75 kΩ; and if the wiper is half round the strip, the two resistance values are both 50 kΩ.

It is also possible to have a situation whereby a logarithmic (taper) scale is used rather than a linear one. *Figure 17* below shows how the two differ.

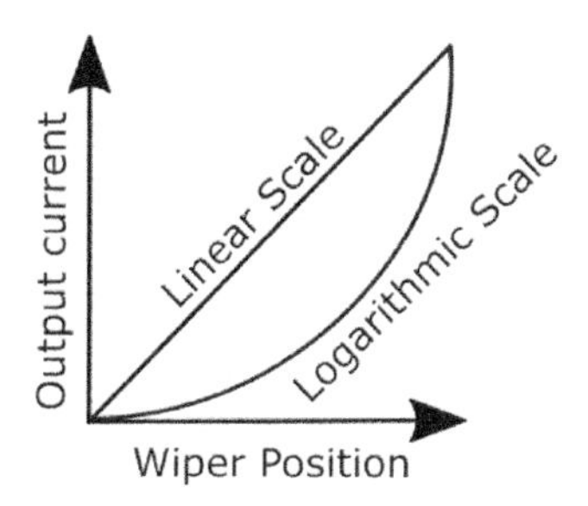

Figure 17 - Linear versus logarithmic

When a logarithmic scale is used, the output current initially increases very slowly as the wiper position is adjusted, but increases at a high rate towards the end of the wiper adjustment.

Logarithmic scales are typically used to control the volume on audio devices.

Photoresistor / Light Dependent Resistor (LDR)

As the name suggests, the amount of resistance provided by an LDR depends on how much light is detected by the LDR. In low light conditions, the resistance of an LDR is very high, but when a torch is shone on it, for example, the resistance drops dramatically to let current flow through the device.

Thermistor

A thermistor is a type of resistor in which the amount of resistance provided by the device varies according to the surrounding temperature.

- At low temperatures, the resistance of a thermistor is high, and only a small amount of current can flow through it.
- At high temperatures, the resistance is low, allowing more current to flow through the device.

As with fixed value resistors, thermistors are colour coded so you can work out the resistance value by looking at the colours. So, for example, a thermistor with a 10 kΩ maximum resistance value has the colours brown, black, and orange as shown in *Figure 18.*

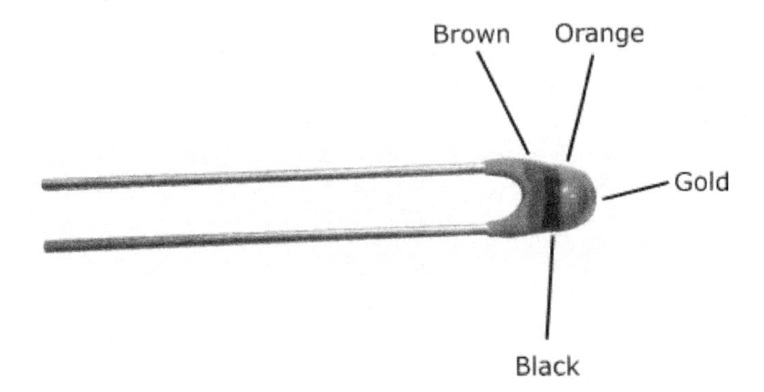

Figure 18 - A 10 kΩ thermistor

Using the following table we can calculate the resistance value.

Colour	Digit 1	Digit 2	Multiplier
Black	0	0	x 10^0

Brown	1	1	x 10^1
Orange	3	3	x 10^3

Table 5 – Colour coding for a thermistor

From the table:

- The first colour is brown, which gives us a value of 1 (Digit 1 column)
- The second colour is black, which gives us a value of 0 (Digit 2 column)
- The third colour is orange, which gives us a multiplier of 10^3 (Multiplier column)

This produces the following resistance value:

- 10 x 10^3, which equals 10 x 10 x 10 x 10, which comes to 10,000 Ω (or 10 kΩ)

You may have noticed in *Figure 18* on the previous page that the end of the thermistor is labelled as *Gold*. This is because - as with fixed value resistors - thermistors have a tolerance and the colour at the end of the thermistor indicates what that tolerance is. Gold means a tolerance of ±5%, meaning that the true resistance value of our 10 kΩ thermistor is somewhere between 9.5 kΩ and 10.5 kΩ.

Thermistors are typically used in fire alarms.

Schematic Diagram Symbols

The schematic symbols for resistors are shown below.

Component	Symbol	Identifier
Fixed Value Resistor (IEEE)		R, R1, R2, R3, and so on
Fixed Value Resistor (IEC)		
Variable Resistor / Potentiometer (IEEE) 3 Terminal		
Variable Resistor / Potentiometer (IEC) 3 Terminal		
Variable Resistor / Rheostat (IEEE) 2 Terminal		

Variable Resistor / Rheostat (IEC) 2 Terminal		
Photoresistor / Light Dependent Resistor (LDR) (IEC)	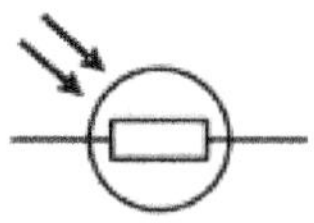	
Photoresistor / Light Dependent Resistor (LDR) (IEEE)	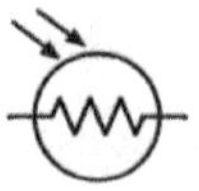	
Thermistor	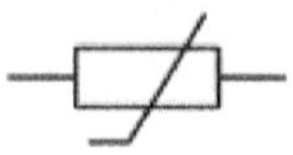	

Table 6 – Resistor schematic symbols

Where there is a choice, in this book we use the IEEE symbols.

Chapter 5

Capacitors

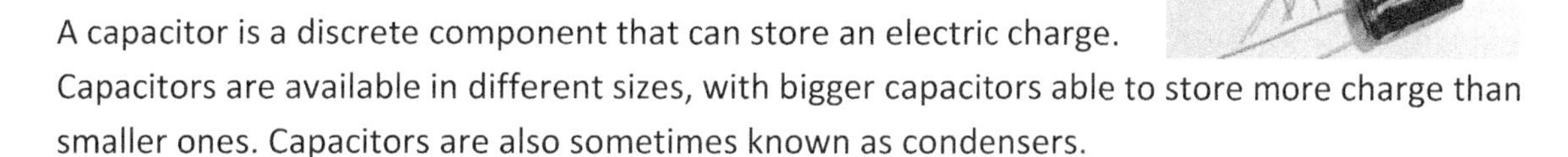

A capacitor is a discrete component that can store an electric charge. Capacitors are available in different sizes, with bigger capacitors able to store more charge than smaller ones. Capacitors are also sometimes known as condensers.

Capacitors are used in lots of electronic equipment, including televisions, radios and cameras. When your camera uses a flash, it's a capacitor that is used to store the electric charge which enables the flash to happen.

A capacitor is similar to a battery in that it is able to store electrical energy. However, in a battery the energy is generally released quite slowing, perhaps over a period of days, weeks or even years. Capacitors on the other hand tend to release their energy very quickly - normally within a few seconds or less.

Capacitors get their electrical energy from another energy source in an electric circuit, such as a battery in the case of a digital camera. Adding electrical energy to a capacitor, which can take a few seconds, is known as *charging*; releasing that energy, which happens very quickly, is known as *discharging*.

Charging and Discharging a Capacitor

A capacitor consists of two conductors that allow electric current to flow, separated by an insulator that does not easily allow electric current to flow. The two conductors are known as *plates*, and the insulator is called the *dielectric*.

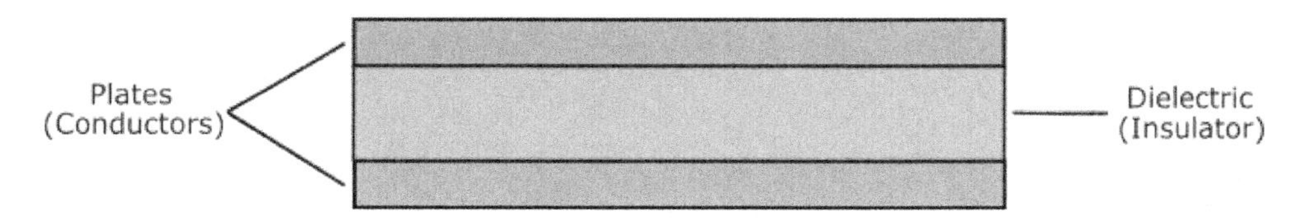

Figure 19 - The parts of a capacitor

To charge a capacitor you connect it to an electric circuit. When power is applied to the circuit, an electric charge gradually builds up on the plates. One plate gains a positive charge and the other gains an equal but opposite (negative) charge. When the power is disconnected from the circuit, the capacitor holds (keeps) its charge. If the charged capacitor is then connected to a device such as a flash bulb in a camera, the charge flows quickly from the capacitor until there is none remaining on the plates.

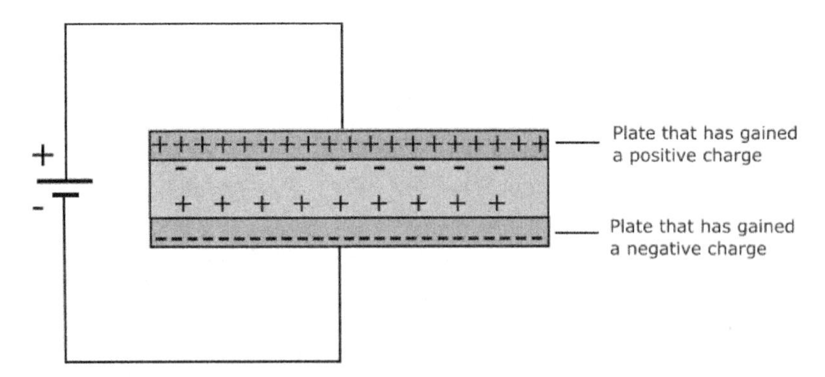

Figure 20 - Charging a capacitor

The positively charged plate and the negatively charged plate are attracted to each other. Separating them requires work to be done against that attraction. The energy required to do this is stored in the electric charges in the two plates as potential energy. This is the energy that is stored when the capacitor is charged, and then released when the capacitor is discharged.

Three things determine how much electric charge a capacitor can store:

- The size of the plates - the bigger the better.
- How close the plates are together - the closer the better.
- How good the insulator (the dielectric) is. Examples are: air, paper, glass, and rubber.

A Bit More About How Capacitors Work

As we have already mentioned, electrons have a negative charge and protons have a positive charge. Opposite charged particles are attracted to each other - electrons like protons, and protons like electrons - but particles with the same charge are repelled by each other - electrons repel electrons, and protons repel protons. This attraction and repulsion is known as electromagnetic force, and it causes all particles to be surrounded by an electric field.

Electric current needs a conductor in order to flow. An electric field, on the other hand, does not. This phenomenon is what capacitors use, whereby the electric field created by the particles in one plate extends across the dielectric and causes particles in the second plate to be attracted or repelled.

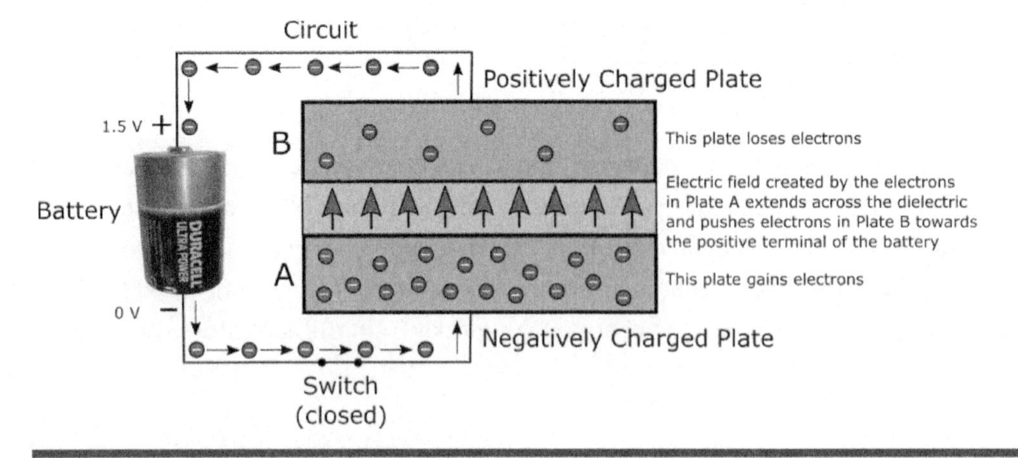

Figure 21 - Charging a capacitor

Looking at *Figure 21* we can see that applying a voltage across the two plates of the capacitor causes electrons to move from one plate to the other. Plate B (which loses electrons) becomes positively charged, while Plate A (which gains electrons) becomes negatively charged. This creates a voltage between the two plates, which increases until the voltage across the two plates is the same as the voltage across the two terminals of the battery - 1.5 V in our case. At this point the capacitor is charged.

If the power supply (the battery) is now removed from the circuit - by opening a switch, for example - the capacitor remains charged and can be used to supply a quick burst of current as and when required in another part of the circuit. When this happens, the capacitor becomes discharged and the plates return to their original state whereby there is no voltage across the two plates, and no current flows.

Electric Field

The electric field that exists between the particles in one plate and those in the other plate stores potential energy. It is the electric field that creates the voltage across the two plates. As the capacitor discharges and the voltage across the plates is reduced, the electric field disappears. By this stage, the energy that the electric field stored will have been converted into other forms of energy such as heat and light.

Below, *Figure 22* shows how the circuit from *Figure 21* can be extended to initially charge the capacitor by closing switch 1, and then discharge it by closing switch 2. When switch 2 is closed, the red LED illuminates for a short period of time before fading. Closing switch 1 again recharges the capacitor and the process can be repeated.

The Resistor, Diode and LED

The resistor is included in the circuit to stop the LED from burning out when the capacitor discharges. The diode is included to ensure that current from the capacitor only flows through the resistor and LED, and does not head off towards the positive battery terminal. The circuit would work without the diode but it is good practice to get into the habit of using them where necessary. Don't worry if you're confused by the use of the diode, we cover their use in the next chapter. We cover LEDs in *Chapter 7* on Page 64.

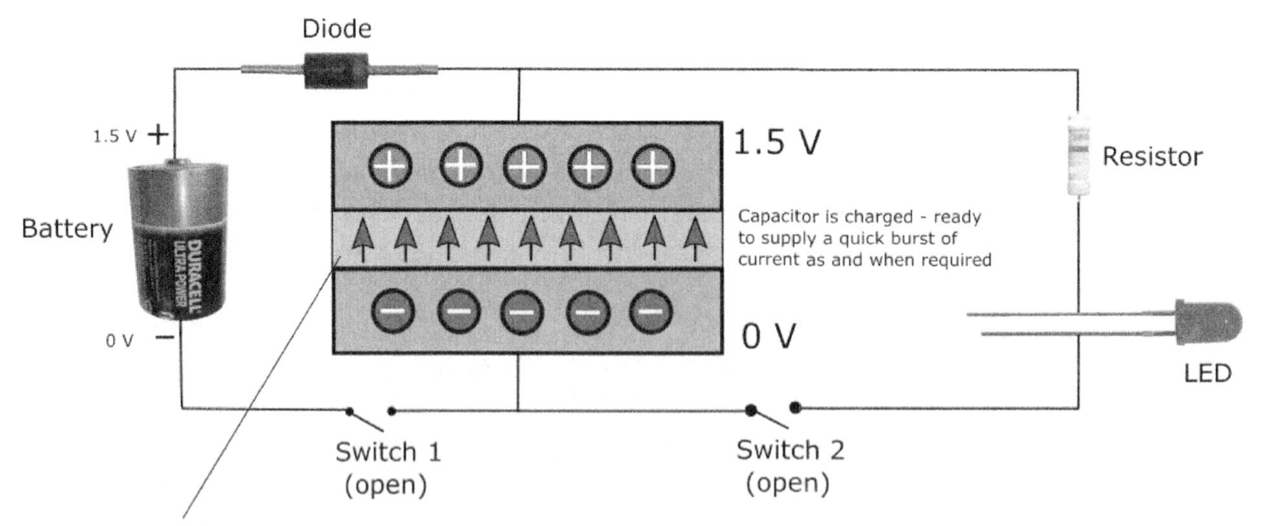

Figure 22 - Circuit to charge and discharge a capacitor

A modified version of the capacitor charge/discharge circuit shown in *Figure 22* above is the subject of Chapter 20 on Page 130.

Measuring Capacitance

The unit of capacitance is the farad (F), named after English physicist Michael Faraday (1791-1867). However, one farad is a lot of capacitance, so it is normal for capacitors to be measured in fractions of a farad such as microfarads (μF), which are millionths of a farad, nanofarads (nF), which are thousand-millionths of a farad, and picofarads (pF), which are million millionths of a farad.

Farads	Microfarads (μF)	Nanofarads (nF)	Picofarads (pF)
0.00000000001	0.00001	0.01	10
0.0000000001	0.0001	0.1	100
0.000000001	0.001	1	1,000
0.00000001	0.01	10	10,000
0.0000001	0.1	100	100,000
0.000001	1	1,000	1,000,000
0.00001	10	10,000	10,000,000
0.0001	100	100,000	100,000,000
0.001	1,000	1,000,000	1,000,000,000

0.01	10,000	10,000,000	10,000,000,000
0.1	100,000	100,000,000	100,000,000,000
1	1,000,000	1,000,000,000	1,000,000,000,000

Table 7 - Capacitance conversion values

Types of Capacitor

Capacitors come in different forms, sizes and styles, but they all contain at least two conducting plates that are separated by a dielectric. Capacitors can have a fixed value, for example, 1000 µF, or they can have a variable value. In variable capacitors, the amount by which the plates overlap can be varied, which allows the amount of charge stored to be varied. Variable capacitors can be used to tune a radio and are often called tuning capacitors.

Fixed capacitors can be separated into two groups:

- Polarized
- Non-polarized

Within each of these two groups there are sub-categories of capacitors:

- Polarized
    - Electrolytic
    - Super-capacitors
- Non-polarized
    - Ceramic
    - Vacuum, air, glass, silicon
    - Film

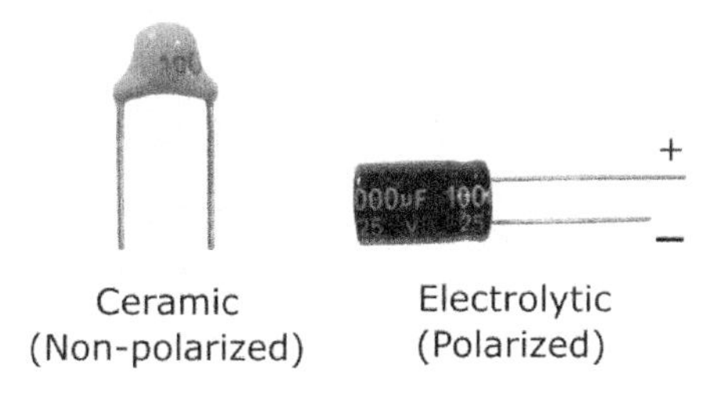

Figure 23 - Electrolytic and ceramic capacitors

Polarized capacitors have one positive and one negative terminal and in order to work must be inserted into a circuit such that the positive terminal is connected to the positive side of the circuit, and the negative terminal is connected to the negative side. If the capacitor has one terminal that is longer than the other, the longer one is the positive terminal. Sometimes there is a '+' and '-' next to the terminals so that you can determine which terminal is which.

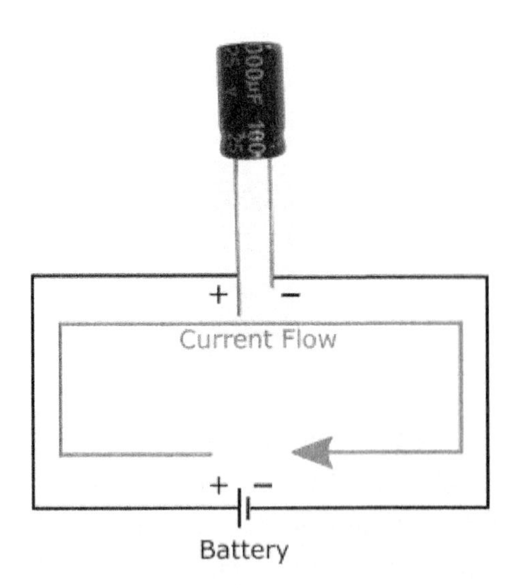

Figure 24 - Correct orientation of an electrolytic capacitor in a circuit

Non-polarized capacitors can be inserted either way round.

Capacitance Value, Voltage Rating and Tolerance

Capacitors often have various letters and numbers written on them that let you identify the capacitance value, voltage rating and tolerance of the capacitor.

Note

Not all capacitors contain sufficient information printed on them to enable you to identify all the properties of the capacitor. Some capacitors are so small that there simply isn't enough room to contain all of the information. Hopefully though, when you bought your capacitor(s), the manufacturer will have supplied you with a data sheet to provide this information.

Capacitance Values

Identifying the value of a larger capacitor such as an electrolytic capacitor is normally fairly easy because you can read the value written on the side of the component.

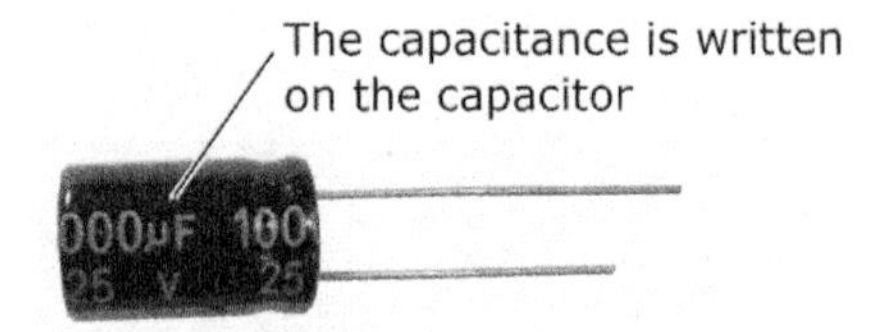

Figure 25 - Reading the capacitance value of an electrolytic capacitor

Reading the capacitance value for a small ceramic capacitor is not quite so straight forward because you have to calculate the value based on the numbers printed on the capacitor.

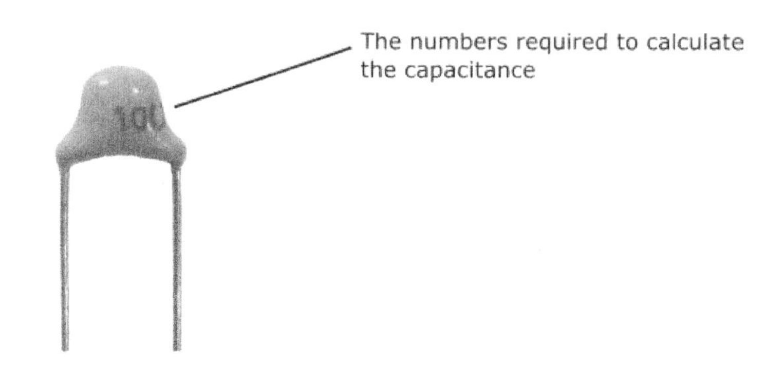

Figure 26 - Reading the capacitance value of a ceramic capacitor

There are 3-digit numbers written on capacitors that enable you to work out their value in picofarads (pF).

1st Digit	2nd Digit	3rd Digit (Multiplier ×)	
0	0	0	×1
1	1	1	×10
2	2	2	×100
3	3	3	×1000
4	4	4	×10000
5	5	5	×100000
6	6	6	×1000000
7	7	7	×10000000
8	8	8	×100000000
9	9	9	×1000000000

Table 8 - Capacitor codes for calculating the value in picofarads (pF)

Example 1

Digit 1 = 1
Digit 2 = 0
Digit 3 = 4

So, 104 is written on the capacitor. Therefore, the value is 10 x 10,000, which equals 100,000 pF, or 100 nanofarads (nF), or 0.1 microfarads (μF).

Example 2

Digit 1 = 1
Digit 2 = 0
Digit 3 = 0

So, 100 is written on the capacitor. Therefore, the value is 10 x 1, which equals 10 pF, or 0.01 nanofarads (nF), or 0.00001 microfarads (μF).

Voltage Rating

This is the maximum voltage the capacitor is designed to handle. For larger capacitors it is written directly on the capacitor, but with smaller capacitors where space is limited, a code is used. Where space is so limited that there isn't even enough room for the code, you'll have to rely on the manufacturer's data sheet for the voltage rating.

Code	Maximum Voltage
1H	50 V
2A	100 V
2T	150 V
2D	200 V
2E	250 V
2G	400 V
2J	630 V

Table 9 - Capacitor voltage rating codes

In all of the experiments in this book we use a battery as the power source, so the maximum voltage any of our capacitors is subjected to is 9 V - well within the values given in *Table 9.*

Tolerance

As with resistors, capacitors have a tolerance rating which means that the value written on the capacitor (or calculated if there isn't room to display the value) is probably not the accurate value. *Table 10*, below, shows the tolerance codes that might be printed on the capacitor.

Code	Tolerance
A	± 0.05 pF
B	± 0.1 pF
C	± 0.25 pF
D	± 0.5 pF
E	± 0.5%
F	± 1%
G	± 2%
H	± 3%

J	± 5%
K	± 10%
L	± 15%
M	± 20%
N	± 30%
P	-0 to +100%
S	-20 to +50%
W	-0 to +200%
X	-20 to +40%
Z	-20 to +80%

Table 10 - EIA codes for capacitor tolerances

Schematic Diagram Symbols

The schematic symbols for capacitors are shown below.

Component	Symbol	Identifier
Non-Polarized Capacitor	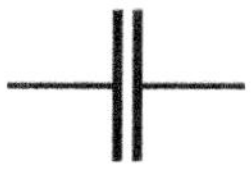	C, C1, C2, C3, and so on
Polarized Capacitor	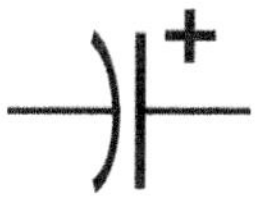	
Polarized Capacitor	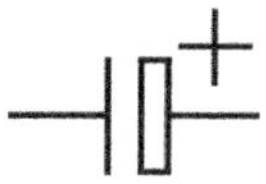	
Variable Capacitor	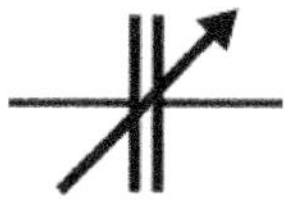	

Table 11 – Capacitor schematic symbols

Chapter 6

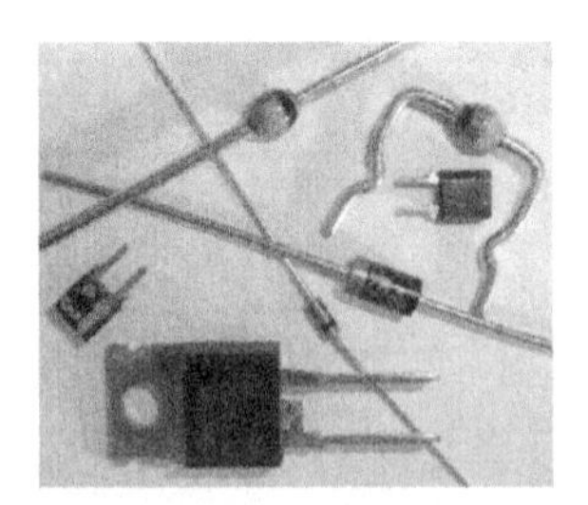

Diodes

A diode is a semiconductor device that passes current in one direction only.

It is a polarized component that has two pins: the anode and the cathode. You can normally identify which is which because there will be some kind of marker on the diode such as a silver or coloured band, which is next to the cathode.

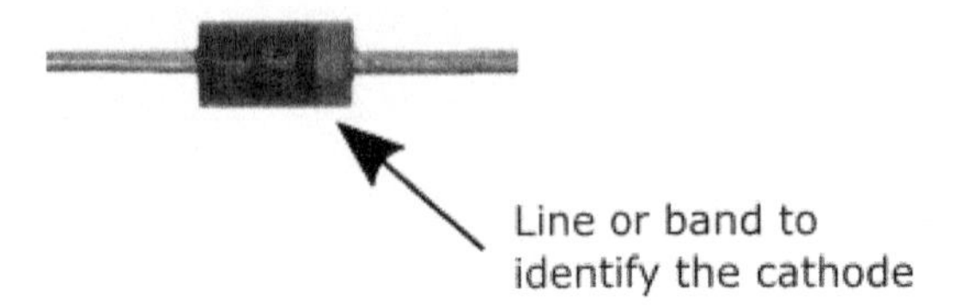

Figure 27 - Identifying the cathode

If the anode is connected to a higher voltage than the cathode (for example, the positive terminal of a battery), current flows through the component (from the anode to the cathode). If the cathode is connected to a higher voltage than the anode, current does not flow through the component.

Different types of diode are available, which provides the ability for circuits to behave differently depending on current, voltage or frequency.

Semiconductors

A semiconductor is a substance that has a conductivity that lies somewhere between that of most metals and an insulator. Components made of a semiconductor substance - such as silicon - are essential in most electronic circuits.

How Diodes Work

In its pure form, silicon is an insulator, but you can turn it into a semiconductor by *doping*. With doping, you mix a small amount of an impurity into the silicon, which changes its behavior. These impurities are classed as being either *N-Type* or *P-Type*.

- **N-type**: With N-Type doping, a small quantity of phosphorus or arsenic is added to the pure silicon. This causes some of the impurity's electrons to break free from their atoms, which are then able to move around within the silicon. As electrons have a negative charge, this causes this part of the diode to also have a negative charge. The *N* in N-Type stands for negative, and this side of the diode is the cathode.

- **P-Type**: With P-Type doping, a small quantity of boron or gallium is added to the pure silicon. The result of doing this is to cause *holes* to be created in the silicon's structure, where there is an absence of electrons, which creates a positive charge within the silicon. The *P* in P-Type stands for positive, and this side of the diode is the anode.

Holes

If a silicon atom loses an electron, a hole is left behind where the electron used to be.

In a diode there are two doped layers: one N-Type and one P-Type. Where these two layers meet, a PN junction is formed. At this junction, the two materials cancel each other out and so a very thin layer is formed that is neither positively nor negatively charged. This layer is called the depletion layer, and no current can flow across it. However, when a voltage is applied across the PN junction, so that the P-Type anode is made positive and the N-Type cathode is made negative, current is able to flow. This is known as forward bias. If, however, the cathode is made positive and the anode negative - in other words the diode is inserted into the circuit the opposite way round - no current is able to flow. This is known as reverse bias.

When the diode is inserted into the circuit like this, no current flows (reverse bias)

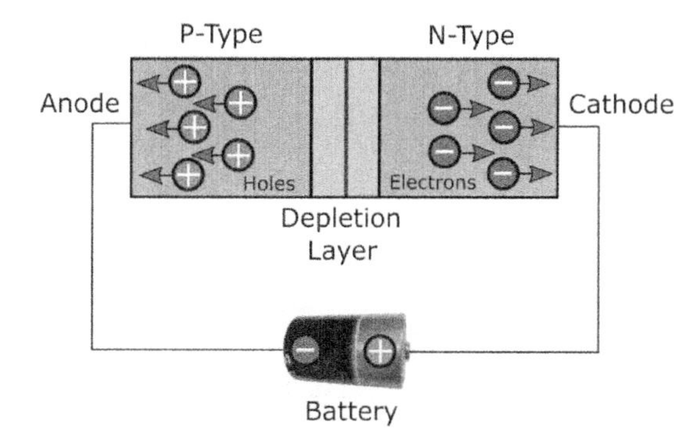

In this circuit, the battery has been connected the opposite way round so that electrons and electric current can flow (forward bias)

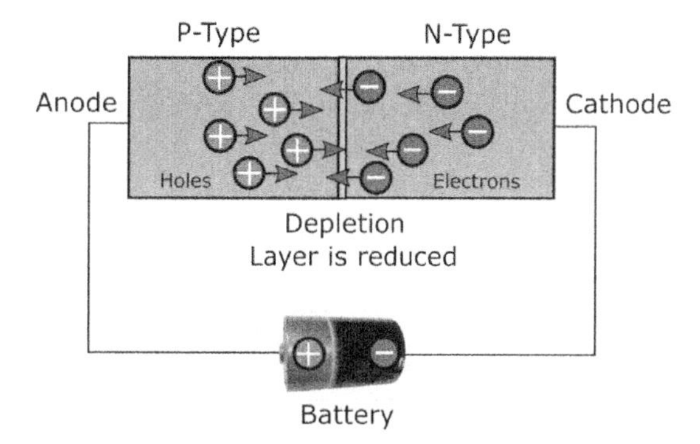

Figure 28 - How a diode works

When current is flowing through a diode, the voltage on the positive pin (anode) is higher than on the negative pin (cathode). This is called the diode's forward voltage drop and needs to be taken into account when trying to determine what size resistor you need to use, for example. We will look at how voltage drop is used in calculations when we build a simple LED circuit in Chapter 13 on Page 93.

Types of Diode

There are several types of diode available, with each type designed to perform a particular function in an electronic circuit.

Regular Diode

This type of diode is designed to stop electricity from flowing in the wrong direction. Whenever a diode is used in this book, it is a regular diode.

Zener Diode

This is a special kind of diode that allows current to flow when the diode is forward biased, but also allows current to flow when the diode is reverse biased so long as the voltage is above a certain value - the breakdown voltage - which is also known as the Zener voltage.

Photo Diode

This type of diode only conducts an electric current when it is exposed to light. A typical application might be a circuit in which a switch only operates when the diode is exposed to light.

Schottky Diode

This type of diode has a low forward voltage drop and can turn on and off very quickly when the breakdown voltage is reached, enabling it to respond very quickly in digital circuits.

Light-Emitting Diode (LED)

The best known type of diode is the light-emitting diode, or LED, which is a diode designed to give off light when a current passes through it.

We use LEDs a lot in the experiments in this book and are worthy of more than just a quick mention in this chapter. As such, we dedicate a whole chapter (Chapter 7 on Page 64) to them.

Schematic Diagram Symbols

The schematic symbols for diodes are shown below.

Component	Symbol	Identifier
Regular Diode		D, D1, D2, D3, and so on
Zener Diode		

Photo Diode		
Schottky Diode		
Light Emitting Diode (LED)		LED, LED1, LED2, and so on

Table 12 – Diode schematic symbols

Chapter 7

Light Emitting Diodes (LEDs)

A light-emitting diode - or LED - is a special type of diode that converts electrical energy into light. LEDs are very popular in many devices that rely on electronics in some way and you will find them in loads of places, for example, coffee machines, cars, telephones, power supplies, and digital clocks. You can think of LEDs as tiny light bulbs that only require a very small amount of current to operate - typically around 20 mA (0.020 A).

LEDs are available in a range of colours and sizes and, as with other diodes, only allow current to flow in one direction. This means that when they are used in an electronic circuit, they need to be inserted in a particular way or they won't work.

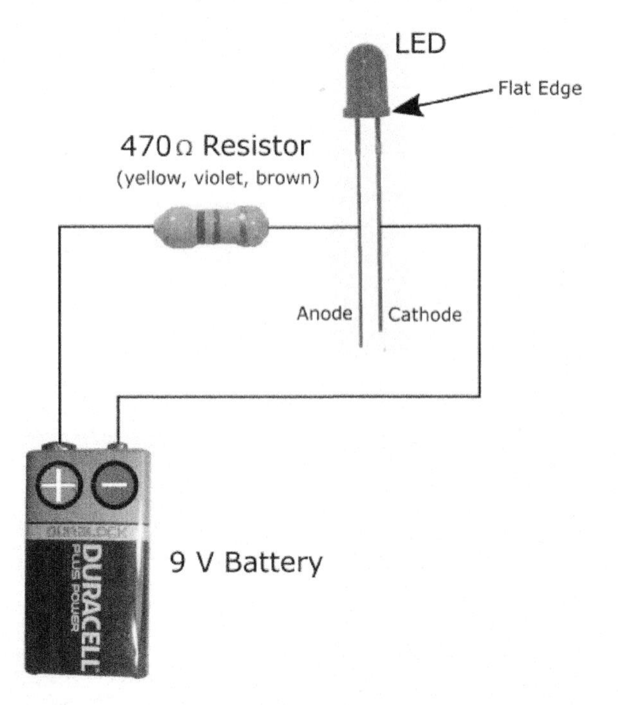

Figure 29 - An LED inserted into an electronic circuit

If you look at the circuit in *Figure 29* you can see that the LED has two pins, one of which is slightly longer than the other. The longer pin is the anode and has to be connected to the positive side of the circuit. The shorter pin is the cathode and has to be connected to the negative side of the circuit. When an LED is connected in this way it is forward biased. Current always flows from the anode to the cathode. There is also a flat edge on the cathode side of the LED, which enables you to identify which pin is which when the LED is inserted into a circuit.

Inserting an LED the wrong way round in a circuit (reverse biased)

If you insert an LED the wrong way round in a circuit (reverse biased) you won't damage it - it just won't work.

Current and Resistance

The more current that flows through the LED, the brighter it shines. Too much current will cause the LED to shine very brightly, but only for a short amount of time before it burns up. Most LEDs require about 20 mA of current to shine brightly, but without presenting any risk of damage to them.

You can control how much current flows through an LED by using a resistor. In Chapter 13 on Page 93 we design and build a simple circuit that shows how to do this (there's also a YouTube video here: https://youtu.be/yQ2-yVXFMeE).

If you use a 9 V battery as the power supply, you will need to use a resistor that is about 400 Ω to sufficiently protect the LED. If you use a battery with a lower voltage, you can get away with using a smaller resistor. If you use a resistor with a higher resistance value than is strictly necessary, the LED will be less bright than it would otherwise be, but it will still work.

To determine what size resistor to use with an LED, use Ohm's Law. Again, in Chapter 13 on Page 93 and the YouTube video, we cover how to do this in some detail.

Heat dissipation

What a resistor does when it is connected in a circuit is to dissipate extra electrical energy (power) as heat energy. This means that if you use a resistor that is perhaps a bit on the big size (in terms of its resistance value) it may well start to get warm because it is having to dissipate a lot of electrical energy as heat in order to bring the current down to a level that is suitable for the LED. If this happens, try using a slightly smaller resistor. Don't go too small though, otherwise you'll run the risk that the resistor doesn't provide enough protection for the LED.

Forward Voltage Drop (V_F)

When current is flowing through an LED, the voltage on the positive leg (anode) is higher than on the negative leg (cathode). This is called the LED's forward voltage drop (V_F) and is the voltage that is "lost" in the LED when it is operated at a certain current.

Voltage drop is a reduction in voltage which occurs as electric current moves through a circuit, and is caused by the internal resistances of the various components in the circuit (resistors, connecting wire, LEDs, and so on).

Electrical energy

> *Electrical energy is a combination of voltage and current, so we can say that a voltage drop also causes a drop in the electrical energy in a circuit. This is important because it is the electrical energy that enables the circuit to do work, for example, illuminate an LED, drive a motor, and so on. If the energy lost in the circuit is too much before an LED is illuminated or a motor is driven, the circuit will fail to do its job properly.*
>
> *Electric power is the rate, per unit time, at which electrical energy is transferred by an electric circuit. The SI unit of power is the watt, which is equivalent to one joule per second. Electric power can be produced by electric generators or electric batteries.*
>
> *To calculate electric power you multiply the voltage by the current.*

Voltage drop is not necessarily a bad thing, and dropping voltage (and therefore electrical energy) across a motor is useful because in doing so, work is being performed by the motor. However, if voltage is dropped across a resistor, for example, this is undesirable because it means that electrical energy is being lost without doing anything useful. For a resistor, the electrical energy is lost in the form of heat energy (see the *Heat dissipation* note earlier in this chapter).

Voltage drop in direct-current (DC) circuits is slightly different to that in alternating-current (AC) circuits. As we're only interested in DC circuits in this book, we will ignore voltage drop in AC circuits.

Voltage drop in direct-current circuits

Consider a circuit that is powered by a 9 V battery, which has one resistor and two LEDs all connected in series as shown below in *Figure 30.*

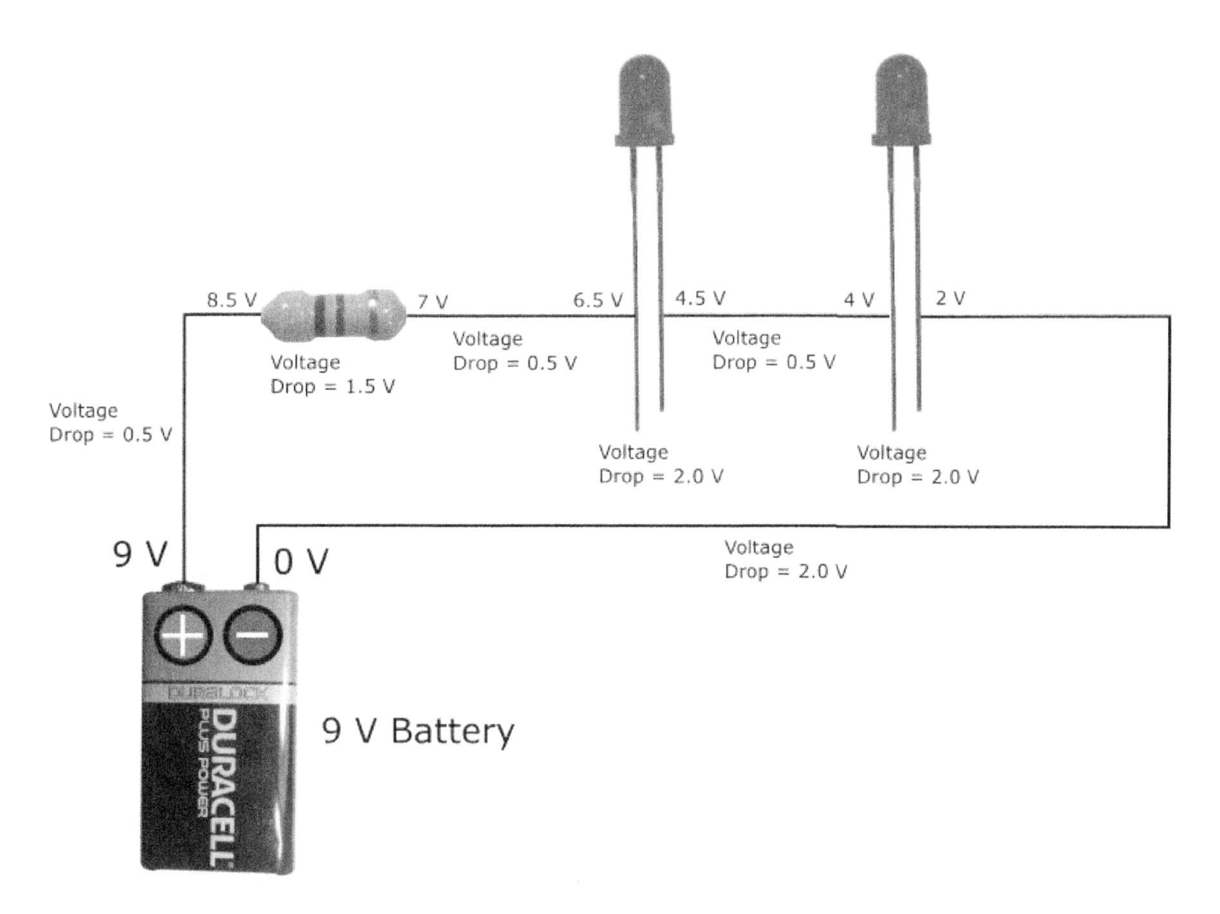

Figure 30 - "Sample" voltage drops in different parts of a circuit

The battery, the conductors (wires), the resistor, and the LEDs (the load) all have resistance and all use and dissipate supplied electrical energy to some degree. Their physical characteristics determine how much energy they use or dissipate. For example, the resistance of a wire depends upon the wire's length, its cross-sectional area, the type of material from which it is made, and its operating temperature.

> *Note*
>
> *The voltage drop values given in Figure 30 are purely intended to provide an example of how voltage is gradually dropped as current passes through a circuit.*

If the voltage between the battery and the resistor is measured, the voltage potential (how much voltage is available) at the first resistor will be slightly less than 9 V. This is because as the current passes through the wire connecting the battery and the resistor, some of the electrical energy is lost due to the resistance of the wire, which means the energy is unavailable for the load (the LEDs). As the current passes through the resistor, more electrical energy is lost due to the resistance of the resistor. More electrical energy is lost as the current heads towards the first LED. As the current passes through the LED, more electrical energy is lost, but this time mainly in the form of light energy. This is good though because this is the purpose of the circuit -

to illuminate this, and the second LED. The process of electrical energy being lost continues throughout the rest of the circuit until we finally get to the negative terminal of the battery at which point the voltage is 0 V.

> *The larger the resistor, the more energy is lost*
>
> *The larger the resistor, the more energy is used by that resistor, and the bigger the voltage drop across that resistor.*

Ohm's Law can be used to verify voltage drop. In a DC circuit, voltage is equal to the current multiplied by the resistance (V = I R). So, a 1 kΩ resistor will produce a larger voltage drop than a 47 Ω resistor, for a given current. Also, Kirchhoff's circuit laws state that in any DC circuit, the sum of the voltage drops across each part of the circuit is equal to the supply voltage. So, for the circuit shown in *Figure 30* on the previous page, the sum of all the voltage drops is equal to the supply voltage of 9 V.

Types of LED

We've already mentioned that LEDs come in different colours (red, green orange, yellow, and so on), but there are also flashing LEDs and RGB LEDs.

Flashing LEDs

Flashing LEDs, as the name suggests, flash when they are connected into a circuit. These LEDs contain a tiny integrated circuit (IC) that allows the LED to flash without requiring the use of an external IC such as a 555 Timer.

Figure 31 - A flashing LED

RGB (Red-Green-Blue) LEDs

RGB LEDs are even more impressive because they are able to glow in a range of colours - not just red, green and blue. RGB LEDs contain three LEDs (red, green and blue) in one package and, by mixing light from the three LEDs, you are able to produce other colours.

RGB LEDs normally have four pins - one for each colour plus a common. Depending on the LED, the common pin might be the anode or it might be the cathode. If you are including an RGB LED in a circuit, you need to know whether the common pin is the anode or the cathode, as well as the colours represented by the other pins. The RGB LED shown below in *Figure 32* is one I have in my electronics kit. Here, the common is the cathode, which means the red, green and blue pins are all anodes.

Figure 32 - An RGB LED

RGB LEDs are different to other LEDs in that you cannot simply connect them to a battery (via a resistor of course) in order to get them to work. Instead, they require a pulse width modulation (PWM) digital signal as input on the red, green, and blue pins, which can be supplied by a micro controller or a 555 Timer. Depending on the LED, the common pin is then either connected to the positive (for a common anode) or negative (for a common cathode) supply.

How an LED is Constructed

The construction of an LED is quite different to that of a 'normal' diode. With an LED, the top of the device is surrounded by a protective, transparent, epoxy (plastic) case that protects the LED and helps to distribute the light generated by the LED.

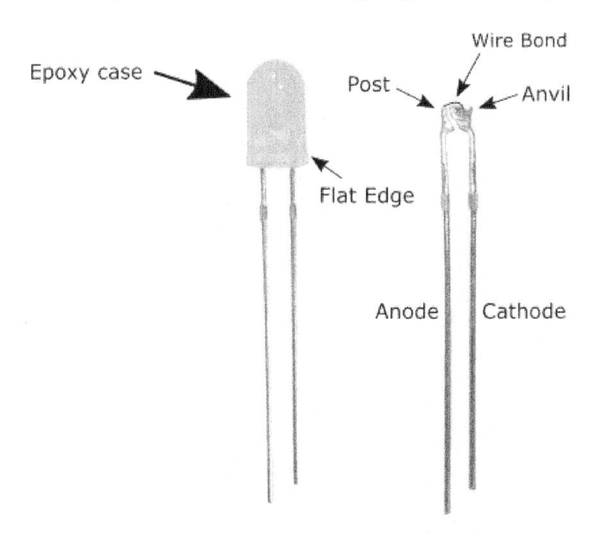

Figure 33 - The construction of an LED

A metal cup (reflective cavity) is placed on the negative pin (Anvil) which holds a semiconductor die, which is a combination of two semiconductor materials – N type and P type - as well as an active region (known as the P-N junction) between them.

The conical shape of the cup reflects the light emitted from the semiconductor die upwards. This is why LEDs appear brighter if you view them from above rather than the side. A wire bond connects the Post and the Anvil.

How an LED Works

The way an LED works is that when it is inserted into a circuit such that it is forward biased, electrons (-) and holes (+) start hopping backwards and forwards across the P-N junction. Whenever an electron finds and occupies a hole in an atom it makes the atom complete, which causes a tiny amount of energy to be released in the form of a photon of light.

The photons are directed upwards towards the domed top of the LED's case, which acts like a lens to concentrate the light.

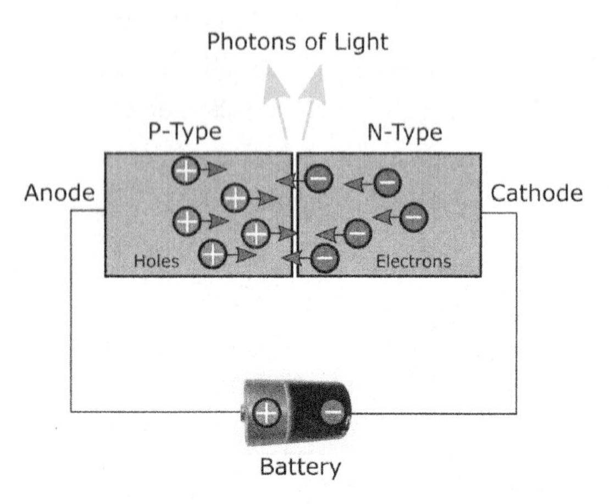

Figure 34 - How an LED works

YouTube Video

There is a YouTube video to accompany this chapter. You can find it here:

https://www.youtube.com/watch?v=npHmj4BzpIM

Chapter 8

Transistors

Of all the electronic components we have looked at so far in this book, the transistor is the one that has had the biggest impact on electronics and the computer industry in general. The transistor is a semiconductor device, and is the building block of modern electronic systems. It can be used as a switch or to amplify electronic signals.

Without transistors, many of the mobile devices we use today would not be mobile at all: laptops, mobile phones, and tablets would be too big to fit in a room in your house, let alone be *portable*.

History

An important part of many electronic circuits is the use of a valve of some sort that can use a small electronic signal to control a much larger signal. Early electrical circuits used vacuum valves to perform this task which, although they worked, were very expensive to run due to the large amount of electricity they consumed. They were also bulky and unreliable because they were prone to overheating.

These days, the transistor performs the role of the vacuum valve but, as it requires very little power to operate, does not overheat like the valves did and therefore is not prone to failure in the same way.

During the early part of the 20th century, progress was made in certain technical and scientific fields that led, in 1947, to the invention of the transistor. In November and December of that year, John Bardeen and Walter Brattain at AT&T's Bell Labs performed a series of experiments and observed that when two gold point contacts were applied to a crystal of germanium, a signal was produced that had an output power greater than the input. This was the break through moment for the transistor and, over the next few months, Bardeen and Brattain, along with William Shockley (their boss), worked on expanding the knowledge of semiconductors. The three men are jointly credited with being the inventors of the point-contact transistor, which was the first type of solid-state electronic transistor.

During the 1960s and 70s people became familiar with transistors due to the popularity of transistor radios, which were much smaller than previous radios. Today, the transistor is the most important component in most electronic circuits, with some ICs containing tens of millions of transistors.

Figure 35 - 1948 photo of (from left) John Bardeen (May 23, 1908 - January 30, 1991), William Shockley (February 13, 1910 - August 12, 1989) and Walter Brattain (February 10, 1902 - October 13, 1987) - the inventors of the transistor

Basic Operation of a Transistor

A transistor is able to use a small signal that is applied between one pair of its pins (for example, the Base and Emitter) to control a larger signal across another pair of pins (for example, the Collector and Emitter). This property is referred to as *gain*, and it occurs when a stronger output signal (voltage or current) is produced that is proportional to a weaker input signal. In other words, the signal is amplified. Alternatively, a transistor can be used to turn current on or off in a circuit as an electrically controlled switch, where the amount of current is determined by other circuit components, for example, resistors.

There are two types of transistors:

- Bipolar Junction Transistor (BJT)
- Field-Effect Transistor (FET)

A bipolar junction transistor has pins labeled Base, Collector, and Emitter. A small current at the Base pin - flowing between the Base and the Emitter - can control or switch a much larger current flowing between the Collector and Emitter pins. For a field-effect transistor, the pins are labeled Gate, Source, and Drain, and a voltage at the Gate can control a current between Source and Drain. In this book we only use bipolar junction transistors, so any reference to transistors infers bipolar junction transistors.

How a Transistor is Constructed

In Chapter 6 on Page 60 we looked at diodes, which are semiconductor devices containing a single P-N junction and two leads: one connected to the P Type semiconductor, and one connected to the N-Type semiconductor. If you remember, a diode allows current to flow in one direction only.

A transistor is similar to a diode except it has an additional semiconductor connected to one end, which can be either N-Type or P-Type. Therefore, instead of just having one P-N junction, a transistor has two. This gives rise to two possible semiconductor combinations:

- PNP transistors, and
- NPN transistors.

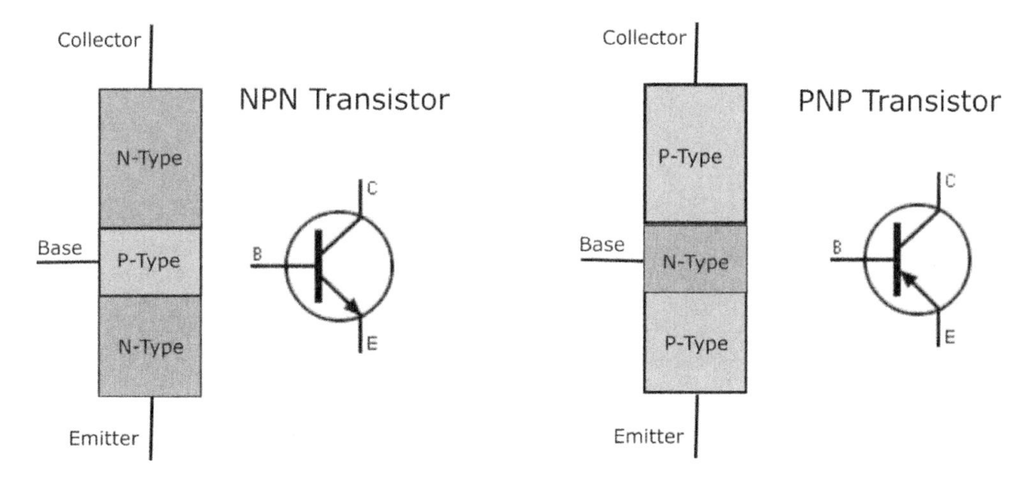

Figure 36 - NPN and PNP transistors

NPN and PNP Transistors

A transistor, whether it is an NPN or a PNP transistor, will typically look like the image shown in *Figure 37*.

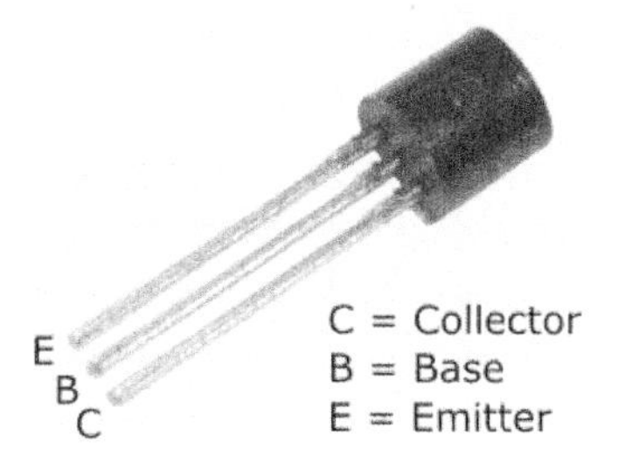

Figure 37 - A typical transistor

> *Which pin is which?*
>
> *The way that the pins are labeled in Figure 37 is not true for all transistors. The best way to identify which pin is which is to refer to the data sheet supplied with the transistor(s) if you can. Table 13, later in this chapter, provides some guidance on pin identification.*

Although NPN and PNP transistors are constructed differently and have to be connected up differently in a circuit in order for them to work, they both do the same thing: they allow for amplification and/or switching. If you design a circuit that uses, say, an NPN transistor, it is quite often easy to change it to use a PNP transistor, as shown in *Figure 38*.

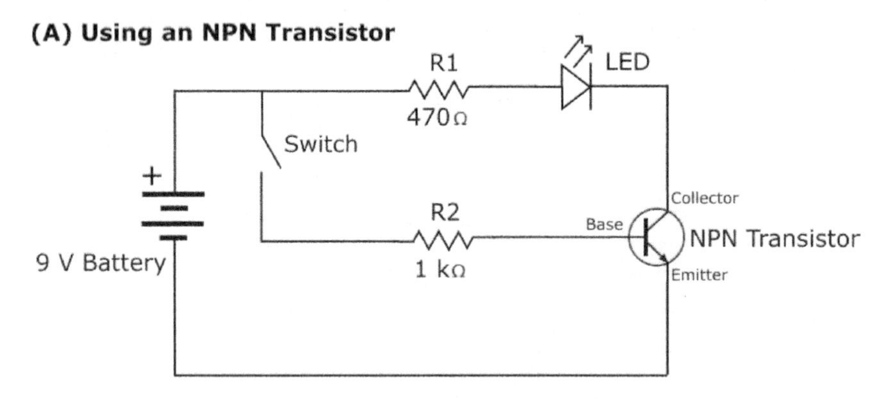

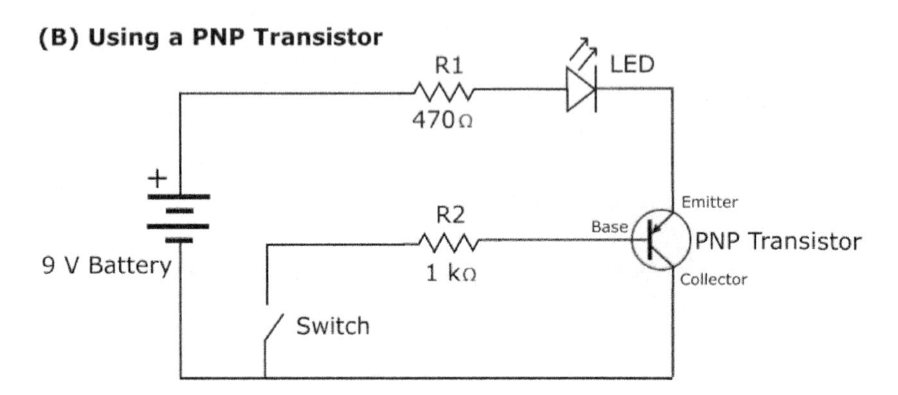

Figure 38 - Switching circuit that uses (A) an NPN transistor, and (B) a PNP transistor

Although the two circuits do the same thing and look very similar, it is important to note the differences in the way that the transistors are connected up.

Due to them being internally constructed differently, the way that voltage and current is allocated within PNP and NPN transistors is different. An NPN transistor needs a positive voltage at both the Collector and the Base, whereas a PNP transistor needs a positive voltage at the Emitter and a more negative voltage (than at the Emitter) at the Base.

Since voltage allocation is different between the two types of transistor, the way current is used to switch them on is also different. An NPN transistor is switched on when sufficient current is supplied to the Base. Therefore, the Base of an NPN transistor must be connected to a positive voltage for current to flow into the Base. A PNP transistor is the opposite; current flows out of the Base by giving it a more negative (a lower) voltage than is supplied to the Emitter.

Another thing that is different between NPN and PNP transistors is that the direction of their output current is different. In an NPN transistor, output current flows from the Collector to the Emitter, whereas in a PNP transistor, output current flows from the Emitter to the Collector.

How NPN and PNP Transistors are Switched On and Off

NPN Transistor

As you increase current to the Base, the transistor is turned on more and more until it conducts fully from the Collector to Emitter.

As you decrease current to the Base, the transistor turns on less and less, until the current is so low that the transistor no longer conducts across the Collector to the Emitter, and switches off.

PNP Transistor

As current flows from the Base toward the negative battery terminal, the transistor is on and conducts current across the Emitter and Collector.

Identifying the Pins

Although transistors come in different shapes and sizes, they all have three pins. Identifying which is which is not always easy though, so the best course of action when you need to identify them is to refer to the data sheet supplied with the transistor if you can. If you don't have the data sheet available, *Table 13*, below, may be useful (hopefully).

Pin Arrangement Guidance for Various Bipolar Transistors	
123 Pin Identification	
NPN (C, B, E)	
Transistor Name	Pins 1 2 3
BC 546, 547, 548, 549, 550, BC 337, AC 187	CBE

TIP 120, 121, 122	BCE
BD 139	ECB
BF 494, 495	CEB
C2570	BEC
C1730	ECB
BD 233	ECB
BD 647, 677	BCE
D882 / 2SD882	ECB
D313 / MJE 13005	BCE
D44VH10	BCE
SF 245	CBE
BUF 742	BCE

PNP

Transistor Name	Pins 1 2 3
2N 2222A, 2N 3904	EBC
TIP125, 126, 127	EBC
BD140	ECB
MPSA 92, 42, 44	EBC
BC 636	ECB
SK / CK / BEL 100P	EBC
AC188	EBC
BC 556 B	CBE

BC 557, 558	EBC

Table 13 - Guidance on pin arrangement for various transistors

Schematic Diagram Symbols

The schematic symbols for transistors are shown below.

Component	Symbol	Identifier
NPN Bipolar Transistor	C B E	Q, Q1, Q2, and so on
PNP Bipolar Transistor	C B E	

Table 14 - Transistor schematic symbols

YouTube Video

There is a YouTube video to accompany this chapter. You can find it here:

https://youtu.be/FHX398piPf0

Chapter 9

Integrated Circuits

So far in this book we have looked at individual electronic components such as transistors, diodes and resistors. These are known as discrete components and by using them together we are able to build electronic circuits that do different things. It is possible, however, to miniaturize these discrete components and combine lots of them into a single electronic component called an integrated circuit (IC) that can perform a particular task or tasks. ICs - also called silicon chips or just chips - can be complicated, as is the case with a random access memory (RAM) chip, or they can be relatively simple, as is the case with a 555 Timer (the subject of the next chapter). ICs can themselves be combined with other ICs and discrete components to create complete electronic systems, for example, a computer's system (mother) board.

Invented in about 1960, the IC revolutionized the computer industry, enabling computer systems that were previously enormous to be made much smaller. This allowed more computing power to be packaged into a smaller container.

An IC is built from a single piece of silicon crystal with all the individual components embedded directly into the crystal. The number of transistors contained in a modern-day IC can be truly staggering, with some microprocessors containing billions of them. At the other end of the scale, the 555 Timer contains about 25 transistors, along with a few other components.

How Integrated Circuits are Packaged

Although there are a few ways in which ICs can be packaged, the most common package - and the only one we examine in this book - is the dual inline package (DIP).

Figure 39 - A 555 Timer packaged in a DIP

Looking at *Figure 39* we can see that a DIP IC comprises a rectangular case and two rows of pins. The IC die itself is housed within the case, which is normally made of some kind of plastic resin, but can be made of ceramic.

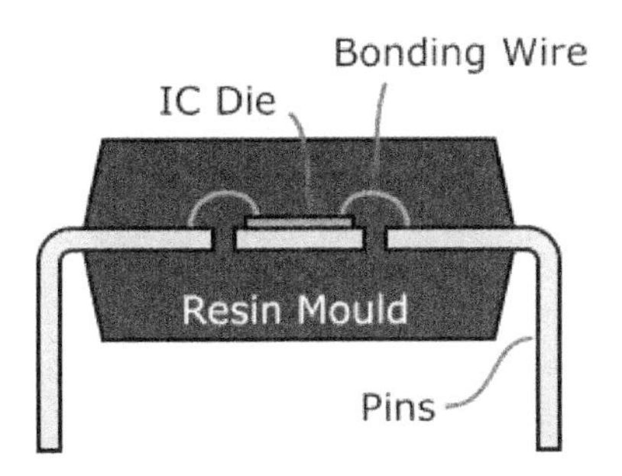

Figure 40 - Side view of a DIP IC (Original image by Inductiveload - Own work, Public Domain, https://commons.wikimedia.org/w/index.php?curid=5892911)

The pins protrude from the longer sides of the package along the seam and are bent downwards at 90 degrees (or slightly less).

Inside the package, the pins are embedded in the lower half. At the center of the package is a rectangular space into which the IC die is cemented. Ultra-fine bond wires are used to connect the die and the pins. If a single bond wire breaks or detaches, the entire IC may become useless. The top of the package covers this delicate assembly, protecting it from damage or contamination by foreign materials.

Pin Identification

All the pins in a DIP are numbered. In order to be able to identify which pin is which, there is normally a mark of some sort next to pin 1.

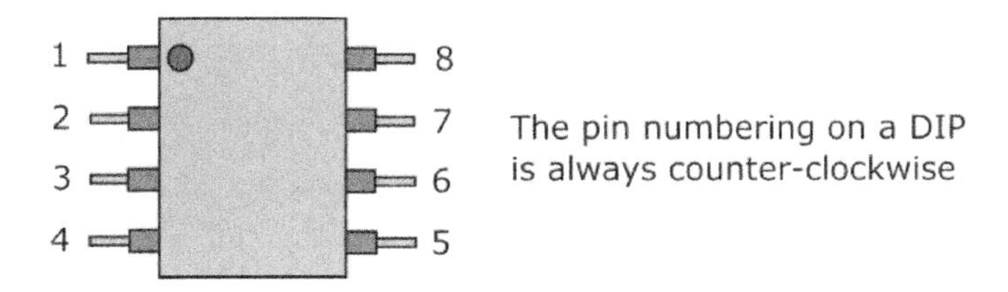

Figure 41 - Pin numbering on an 8-pin DIP

Showing ICs in a Schematic Diagram

Although the pin numbering on a DIP is as shown in *Figure 41*, when an IC is shown in a schematic diagram the positioning of the pins may well be different. For example, when a 555 Timer is shown in a schematic diagram the pins are (almost) always shown in a particular way, and any unused pins are not shown at all. We'll see more of this when we look at the 555 Timer in the next chapter.

Chapter 10

The 555 Timer

Introduced in 1971, and with an estimated 1 billion (or more) being produced every year, the 555 Timer is quite possibly the most popular IC in the world.

The circuit contained within the 555 Timer chip is called a multivibrator, which is used to implement a variety of simple two-state devices such as:

- Relaxation oscillators - a circuit that produces a repetitive output signal, such as a square wave. This could be used to cause a warning light to flash on and off, for example. When used in this way, the 555 Timer is said to be in Astable mode.

  Astable means *no stable state.*

- Timers - a circuit enabling a light to be switched on for a certain amount of time, for example. When used in this way, the 555 Timer is said to be in Monostable mode.

  Monostable means *one stable state.*

- Flip-flops - a circuit that has two stable states, for example, on receipt of a trigger, the circuit can permanently produce an output of +5 V until a second trigger causes the output to go to 0 V. It will then stay at 0 V until a third trigger is received, which will cause the output to go back to +5 V, and so on. Flip-flops are one of the fundamental building blocks of digital electronics systems used in computers and communications systems. When used in this way, the 555 Timer is said to be in Bistable mode.

  Bistable means *two stable states.*

The output from a 555 Timer typically looks like that shown below in *Figure 42*. This is a digital (rectangular) signal, which can be either off (0 V), or on (say 3 V). How long the signal is on or off for is determined by the timing mechanism inside the 555 Timer's circuit along with the circuitry connected to the timer.

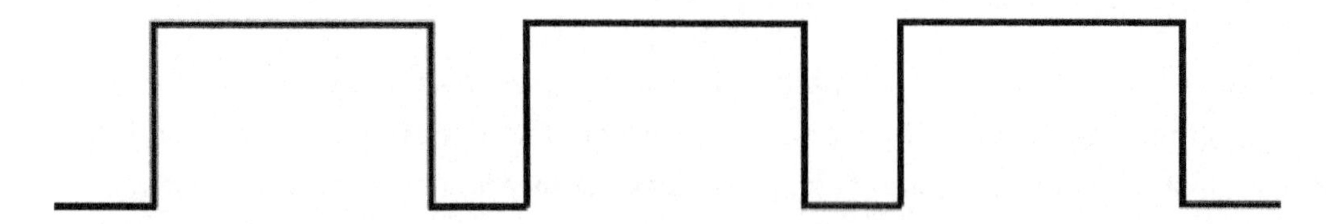

Figure 42 - Typical output from a 555 Timer

The 555 Timer is an analog-digital device. The output, as shown above, is a digital signal, but the input to the chip is an analog signal produced by a resistor-capacitor (RC) circuit. It is this circuit that determines how long the output signal is high or low, which can be varied by using different value resistors and capacitors.

> *Some devices require a digital signal as input*
>
> *Some devices such as LEDs work quite happily when a continuous DC signal is applied to them. Other devices, however, such as a servo motor or a piezo buzzer require a digital signal as input.*

555 Timer Pins

The pins on the 555 Timer are as shown below in *Figure 43*.

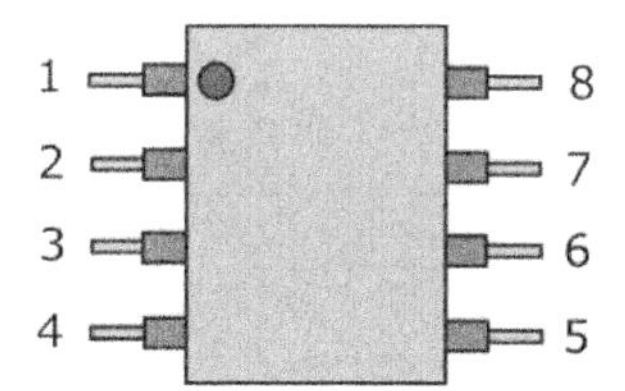

Figure 43 - Pin numbering on the 555 Timer

They are used as described in *Table 15*, below.

555 Pin#	Pin name	Pin purpose
1	GND	**Ground supply:** This is the ground reference voltage (zero volts). You connect this pin to the negative terminal of the battery.
2	TRIG	**Trigger:** The OUT pin (pin 3) goes high and a timing interval starts when the voltage on this (TRIG) pin is kept at a low voltage. The *low* voltage needs to be less than 1/2 of the CTRL voltage, which is typically 1/3 of the supply voltage applied to pin 8 (Vcc). *(Vcc stands for Voltage Common Collector.)*
3	OUT	**Output:** This is where the digital output signal leaves the chip. The output is either very low (almost 0 V) or close to the supply voltage (Vcc) applied to pin 8. How long the output signal is high or low depends on the connections made to pins 2, 4, 5, 6, and 7.
4	RESET	**Reset:** A timing interval may be reset by driving this input to GND, but the timing does not begin again until RESET rises above approximately 0.7 volts

		(triggered by pin 2). In order for the 555 Timer to operate, this pin must be connected to the supply voltage (Vcc). This pin overrides TRIG, which in turn overrides THR (unless you are using an LM555 Timer, in which case THR overrides TRIG).
5	CTRL	**Control:** This pin provides "control" access to the 555 Timer's internal voltage divider. By applying a voltage to this pin you can alter the timing characteristics of the device. However, in most applications, this pin is not used and is connected to ground via a small (0.01 μF) decoupling capacitor that is used to smooth out any fluctuations in the supply voltage that may affect the operation of the timer.
6	THR	**Threshold:** The purpose of this pin is to monitor the voltage across the capacitor that is discharged by pin 7 (DIS). When the voltage reaches 2/3 of the supply voltage (Vcc), the output on pin 3 goes low, ending the timing cycle.
7	DIS	**Discharge:** This pin is used to discharge an external capacitor that is working in conjunction with a resistor to control the timing interval.
8	V_{CC}	**Positive supply:** This pin is connected to the positive supply voltage, which must be in the range 4.5 to 15 V. The 9 V battery supply we use in the circuits in this book is ideal, but you could equally use four 1.5 V AA or AAA batteries to obtain a supply voltage of 6 V.

Table 15 - 555 Timer Pins and Descriptions

How the Pins on a 555 Timer are Normally Shown in a Schematic Diagram

When a 555 Timer is shown in a schematic diagram, the pins are not positioned as shown above in *Figure 43*. Instead they are almost always shown as in *Figure 44* below.

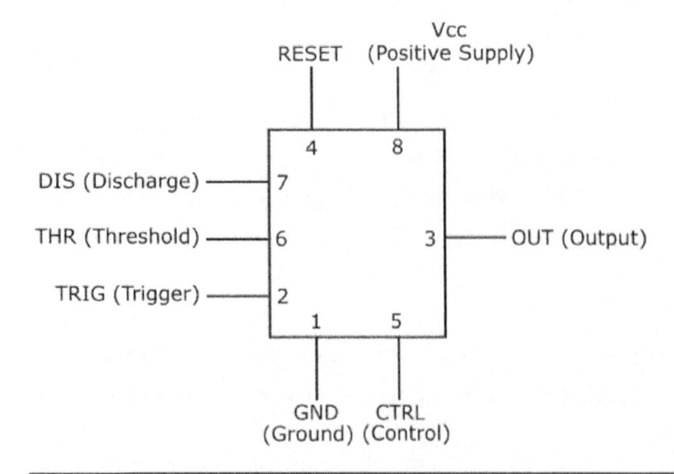

Figure 44 - How the pins of a 555 Timer are normally shown in a schematic diagram

A Look Inside the 555 Timer

Depending on the manufacturer, the 555 Timer contains the equivalent of 25 transistors, about 16 resistors and two diodes. *Figure 45* shows a functional block diagram of the NE555 Timer, which is one of the many 555 Timers on the market.

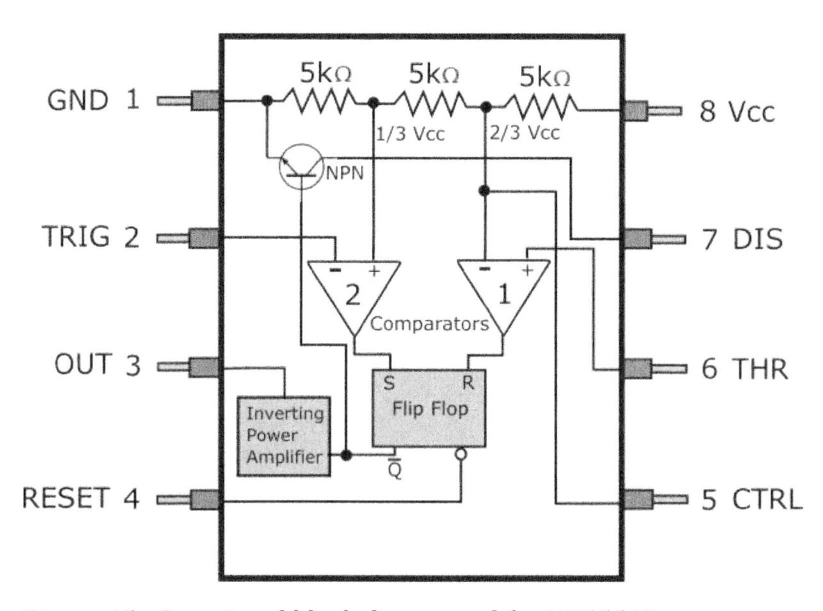

Figure 45 - Functional block diagram of the NE555 Timer

Basic Operation

With reference to *Figure 45*. The three 5k resistors are what give the 555 Timer its name. They create a voltage divider circuit that connects the supply voltage (Vcc) at pin 8 and ground (0 V) at pin 1.

The 555 Timer contains two comparators. A comparator is a circuit that compares two input signals (voltage or current) and outputs a single signal. In the 555 Timer shown above, Comparator 1 is the Threshold Comparator, and Comparator 2 is the Trigger Comparator.

Comparator 1, which provides the Reset input to the SR Flip-flop, compares the threshold voltage with a 2/3 Vcc reference voltage.

Comparator 2, which provides the Set input to the flip-flop, compares the trigger voltage with a 1/3 Vcc reference voltage.

The two comparators produce an output voltage dependent upon the voltage difference at their inputs, which is determined by the charging and discharging action of an externally connected Resistor-Capacitor network (*see note below*).

Note

In order to do something useful, a 555 Timer needs to be connected to an external circuit.

The outputs from both comparators are connected to the two inputs of the SR Flip-flop which in turn produces either a high or low level output at Q based on the states of its inputs. The output from the flip-flop is used to control a high current output switching stage to drive the connected load producing either a high or low voltage level at the output pin.

Chapter 11

Combining Electronics with Software

Although programming is outside of the scope of this book, it is probably worth mentioning how electronics and software co-exist in many electronic systems produced today. In fact, many tasks that used to be performed by using dedicated electronic systems are now done using software. This is a trend that started many years ago, but has become more prevalent in recent years.

Designing and developing a hardware system is a very slow and expensive process, and you need to make sure there are no faults in the system when it goes into production as fixing them will almost certainly be a very costly exercise. Software systems on the other hand can normally be designed and developed more quickly, and errors (or bugs) can be more easily fixed at a later stage.

Companies are often able to buy an "off the shelf" hardware product and then program it to perform a particular task. This is normally quicker and easier (and therefore cheaper) than designing and developing the hardware product from scratch in house.

Adding Computing Power to Electronics Systems

In this chapter we take a quick look at microcontrollers, which are very small computer-like devices contained on a single chip. Microcontrollers contain many of the components you find in larger computers such as desktop PCs (a microprocessor, random access memory, input/output capabilities to communicate with other components, and so on), but don't have the computing and input/output capabilities to be used as complete standalone systems in their own right. They are normally used as embedded systems in devices such as telephones, cars, and kitchen appliances.

Programming

There are three types of programming language:

- Machine code, which is easy for a computer to understand but not for humans. It is written as a series of 1s and 0s, for example, 00000101, which correspond with the high and low voltage level signals used internally by computers.

- Assembly language, which is a bit more difficult for computers to understand, and slightly easier for humans. Assembly languages use mnemonics to represent machine code instructions, for example, MOV (for move) or LD (for load). In order to be understood by a computer, a utility program called an assembler converts assembly language instructions into machine code.

- High-level languages, which need to be compiled or interpreted before a computer can understand them, but are relatively easy for humans to understand because they use instructions that resemble English. Examples of high-level languages are Python, C, Java, BASIC, and Perl.

A programming language can be used to create a program, which is made up of several lines of instructions. It is unusual these days to need to write programs in assembly language or machine code, but in the early days of microcontrollers that was the case. Nowadays, high level languages can normally be used, and it is quite common for customized versions of high-level languages to exist for a particular microcontroller, for example, for the BBC micro:bit, there is a special version of Python that can be used.

Single Board Computers

The capabilities of a microcontroller can be extended by adding components such as a universal serial bus (USB) connector (to connect to a PC) and pin input/output connectors to enable the microcontroller to communicate with other hardware components in a larger system.

A single-board computer (SBC) is a complete computer built on a single circuit board, with microprocessor(s), memory, input/output (I/O) and other features required of a functional computer.

Open-Topped Single-Board Computers Are Nothing New

Although single-board computers are very popular with hobbyists these days, they are nothing new. The first true single-board computer (see the May 1976 issue of Radio-Electronics) called the "dyna-micro" was based on the Intel C8080A, and also used Intel's first EPROM, the C1702A. Another example of an early single-board computer is the 6502-based EMMA, which was produced by LJ Electronics (now LJ Create) in the mid-1980s and sold to schools and colleges.

Figure 46 - EMMA II Micro Computer from the Mid-Eighties (reproduced with kind permission from LJ Create, Norwich, UK)

Part 2 - Designing and Building Electronic Circuits

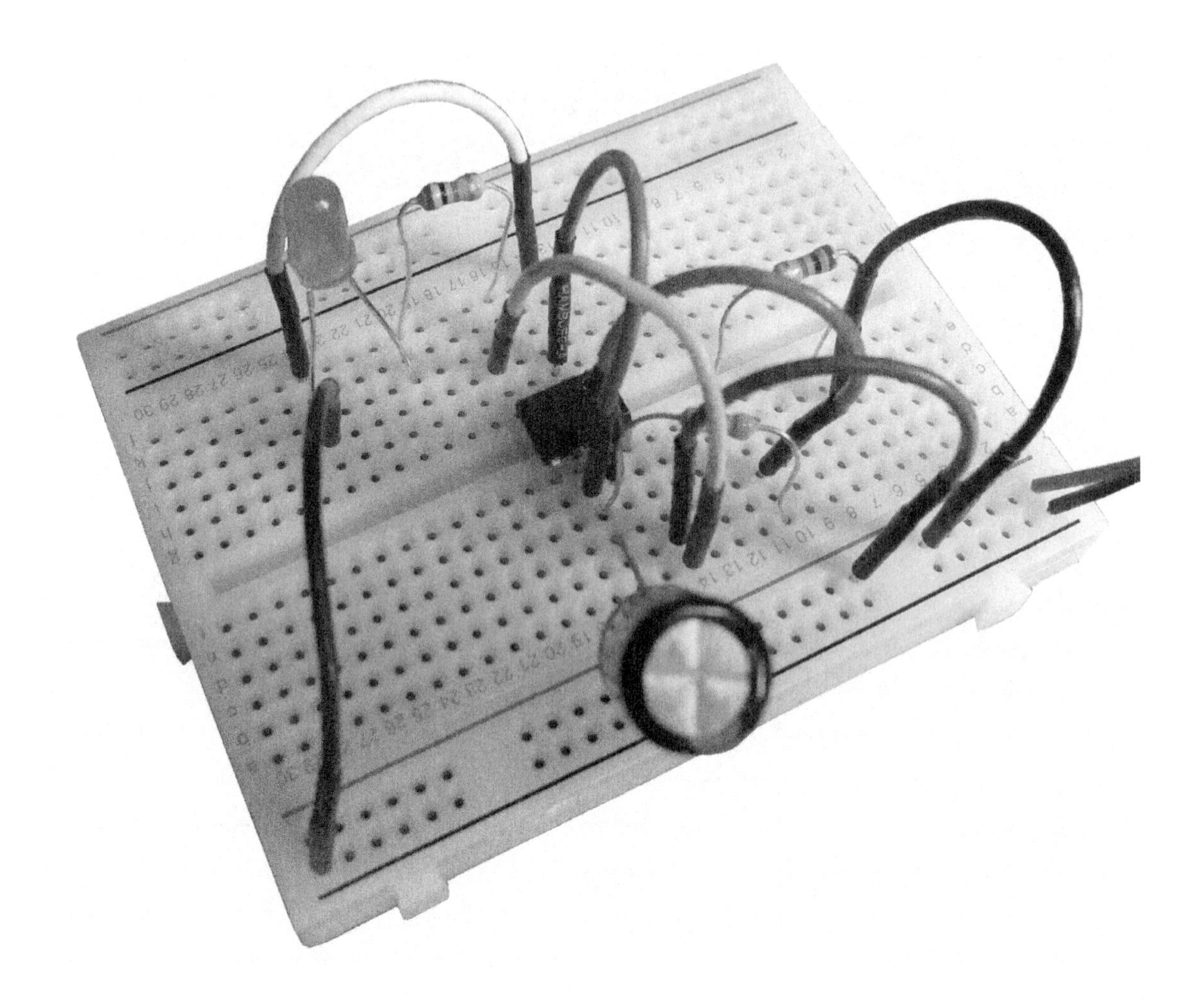

Chapter 12

Using an Electronics Prototyping Breadboard

A breadboard enables you to prototype electronic circuits without the need to solder components in place.

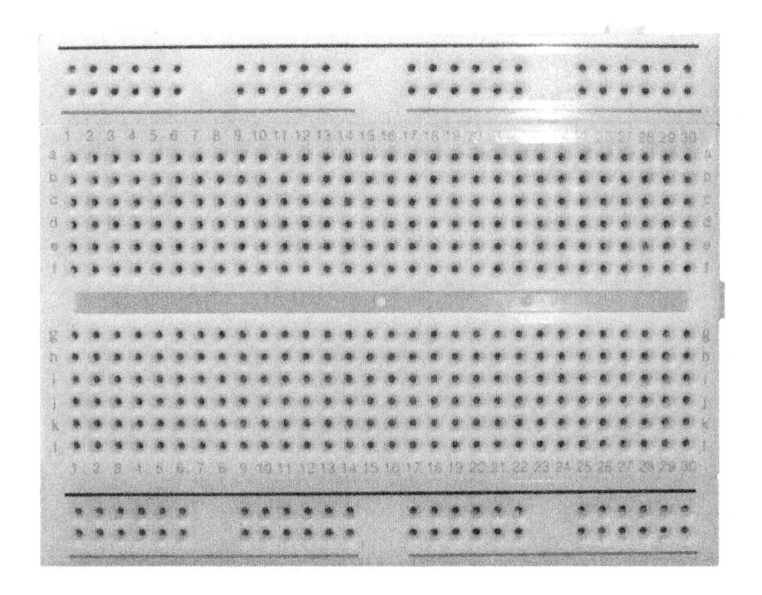

Figure 47 - A prototyping breadboard

Beneath the holes are a series of internally connected rows and columns that enable components to be connected to each other by jumper wires.

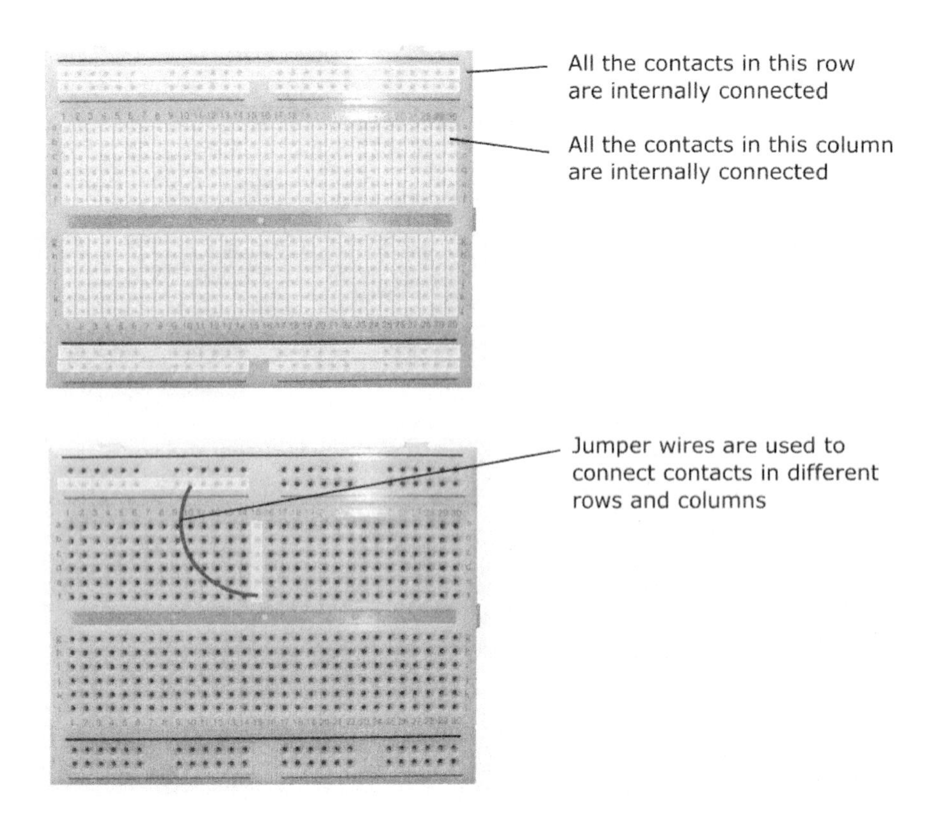

Figure 48 - Internally connected rows and columns

Building Different Circuits

With a breadboard you can insert a component into position and then remove it and put it somewhere else, enabling you to build different circuits.

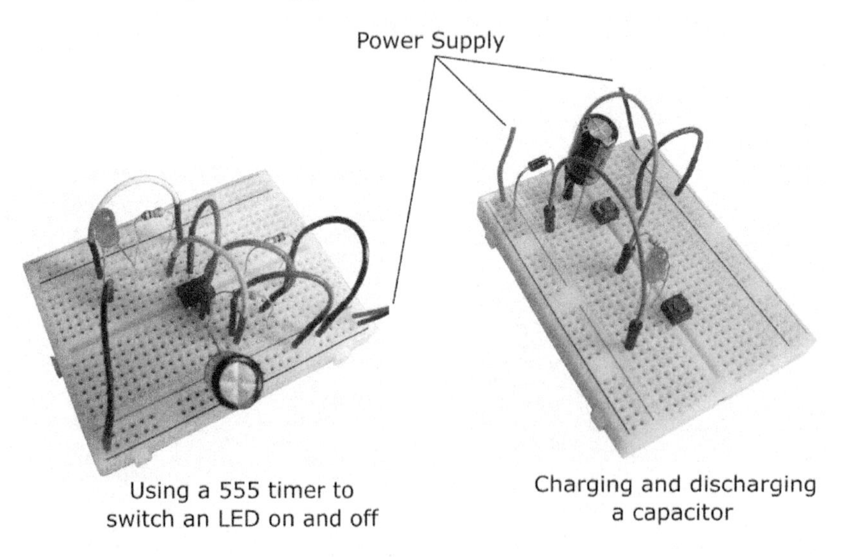

Figure 49 - Building different circuits on a breadboard

Clipping Breadboards Together

Breadboards have little tabs on two edges, and slots on the other two edges, which enable you to clip breadboards together to make a bigger one.

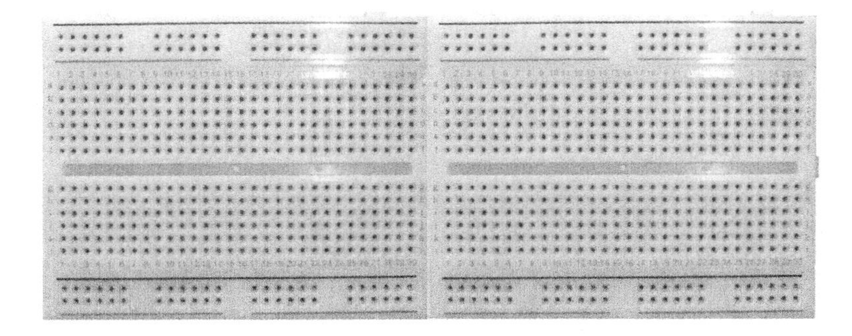

Figure 50 - Clipping breadboards together

Supplying Power to the Breadboard

To supply power to a breadboard you can use:

- a battery
- a single board computer or microcontroller such as an Arduino, BBC micro:bit, or Raspberry Pi
- a benchtop power supply unit
- a purpose-built breadboard power supply

Avoiding Short Circuits

One thing you need to watch out for - especially when you first start building circuits on your breadboard - is short circuits. Consider the two circuits shown below in *Figure 51.*

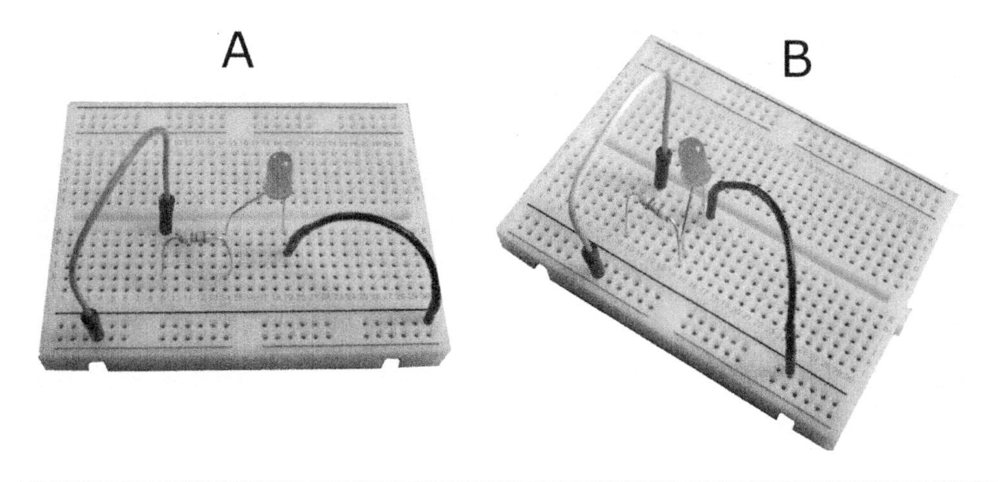

Figure 51 - Two similar circuits designed to illuminate an LED

At first glance these two circuits may look as though they do the same thing, which is to illuminate an LED. Circuit **A** will work (when you apply a power source to it), but circuit **B** won't because a short circuit exists across the LED, as shown below.

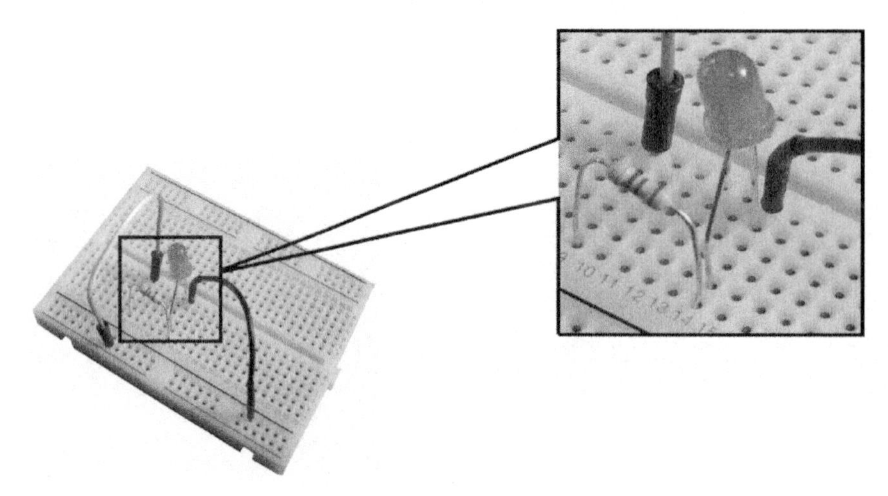

Figure 52 - Short circuit exists across the LED

The LED's anode and cathode are connected to the same column (track) so the current will travel along the breadboard's track rather than through the LED, resulting in the LED not illuminating.

YouTube Video

There is a YouTube video to accompany this chapter. You can find it here:

https://www.youtube.com/watch?v=OfTMcMAeQ-w&t=61s

Chapter 13

Illuminating a Single LED

The purpose of this experiment is to build an electronic circuit to illuminate one LED.

Parts Needed

	Breadboard You build the circuit on this.
DURACELL PLUS POWER	**9 V Battery** This supplies the electrical energy to turn on the LED.
	Battery Clip This lets you connect the battery to the breadboard. **or Battery Box** Instead of using a battery clip, you can insert the 9 V battery into a switched battery box. This means that once you have connected the battery box to the breadboard, you can easily switch the power on and off without needing to disconnect the battery.
	470 Ω Resistor This protects the LED by reducing the amount of current that flows through the LED when the battery is connected/switched on. We use Ohm's Law to show that a 470 Ω resistor is an appropriate size to use for the circuit.

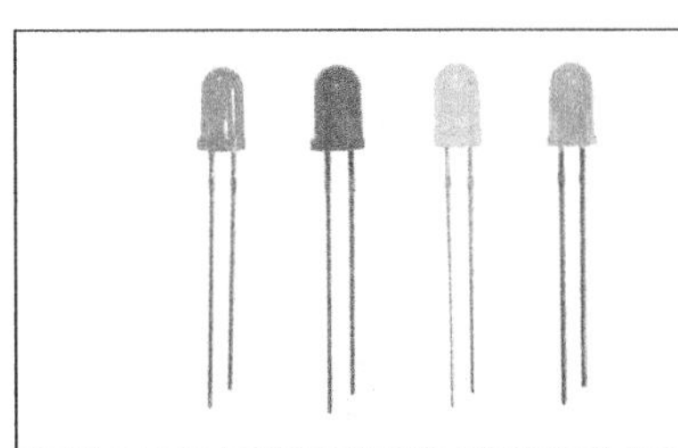	LED You need one LED - it doesn't matter what colour it is, and the range available is quite extensive. Common colours are red, green, yellow and amber.

Table 16 - Parts needed for an LED circuit

Using Ohm's Law to Calculate the Correct Resistor to Use

In *Table 16* we state that the circuit requires a 470 Ω resistor to protect the LED. But how do we know that this is the correct resistor to use? After all, resistors come in a very large range of sizes, so why don't we use a 47 Ω resistor instead? or perhaps a 2 kΩ one? Well, we can use Ohm's Law, which we introduced in Chapter 1 on Page 15, to calculate what size resistor is the most appropriate to use in the circuit.

Ohm's Law states that electric current is proportional to voltage and inversely proportional to resistance. Written as a formula, it can be shown in three ways:

- Voltage (in volts) = Current (in amps) x Resistance (in ohms)
- Current = Voltage / Resistance
- Resistance = Voltage / Current

You need to know two of these values in order to calculate the third. So, for example, if you know the voltage and current, you can calculate the resistance; if you know the resistance and voltage, you can calculate the current; and if you know the resistance and current, you can calculate the voltage.

The only thing we know at the moment is the voltage of the battery, which is 9 V, but we don't know anything else. We have stated that we need to use a 470 Ω resistor, but we don't actually know at this stage whether this is the correct size. In fact, before we go any further, we need a bit more information - we need to know a bit more about the LED we intend to use in the circuit.

Specifications for 5 mm Red, Green, Yellow and Amber LEDs

Table 17, below, gives some guideline specifications for different coloured LEDs.

LED	Current Required to Operate the LED		Forward Voltage Drop
	Maximum (for continuous use)	Maximum (for a very short period)	

	20 mA	30 mA	2.0 to 2.5 V
	20 mA	30 mA	3.0 to 3.6 V
	20 mA	30 mA	1.7 to 2.5 V
	20 mA	30 mA	2.0 to 2.6 V

Table 17 - Guideline specifications for a range of 5 mm LED colours

Caution

You should always check the specifications for the particular LED(s) you plan to use in your projects.

Table 17 gives us some useful information for each colour of LED that we can use to help us calculate the appropriate resistor to use in the circuit.

Let's assume we're going to opt for a red LED.

The LED requires a certain amount of current in order to operate. For normal use - that is, the LED will shine brightly but it will not get damaged - should be no more than 20 mA. This is the value of current we need to use in Ohm's Law.

This means that we now have the two values required (current and voltage) to calculate the third (resistance). However, there is slightly more to it than that. If you look at *Table 17* you'll see that there is another value given - the forward voltage drop. We need to take this into account when entering the voltage value into Ohm's Law. (*If you can't remember what forward voltage drop is, go back and read through the section* Forward Voltage Drop *in* Chapter 7 on Page 64.)

By referring to *Table 17*, for a red LED, 2.0 to 2.5 V is lost when the LED is operated at 20 mA.

The voltage value we need to use is the difference between the supplied voltage (9 V) and the forward voltage drop (we'll use 2 V for this).

So, 9 V - 2 V = 7 V. This is our voltage value for Ohm's Law.

Calculating the Resistor Size

We now have all the necessary values to enter into Ohm's Law to calculate the resistor size:

Voltage = 7 V
Current = 0.020 A

Using Ohm's Law to calculate the resistance value:

Resistance = Voltage / Current

Therefore:

Resistance = 7 / 0.020 = 350 Ω

So, we need to use a 350 Ω resistor ... or should we?

In general, it's a good idea to use a resistor that has a slightly higher value than the calculated value. When we looked at resistors in Chapter 4 on Page 40, we examined the popular E12 Series (*Table 2*). If you refer back to this table you will see that there is a resistor in this series that has a value of 390 Ω, so we could use that. Remember also though that resistors have a tolerance value ... ±10% in the case of the E12 Series. So, a 390 Ω E12 Series resistor could in fact be anywhere between 351 Ω and 429 Ω.

Other resistors have a tolerance of ±5% or even ±20%. In the type of circuits we build in this book, there's no need to worry about using a resistor that has a resistance value a little bit on the high side as all it means in the case of an LED is that the LED will not shine quite as brightly as it would if a lower-value resistor were used.

To make sure that we are well on the safe side as far as damaging the LED is concerned, we will use a 470 Ω resistor in our circuit.

Designing the Circuit

Now that we're confident about all the parts we need we can design the circuit.

Figure 53 (below) shows two versions of the circuit: one where friendly pictorial representations of the components are used, and one where schematic components are used.

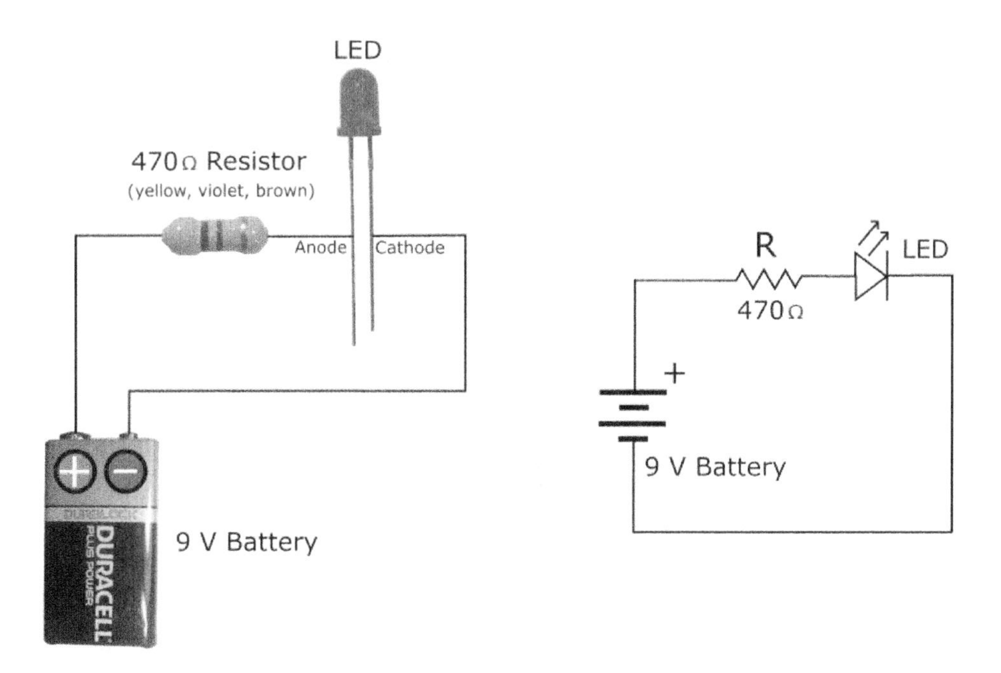

Figure 53 - The LED circuit

Things to note:

- Resistors are not polarized so it doesn't matter which way round they are inserted into a circuit.
- As well as the three colours that determine the resistance value of the resistor (yellow, violet and brown), the gold band indicates that it has a tolerance of ±5%.
- The LED is polarized and will only allow current to flow in one direction. The longer pin (anode) needs to be closer to the positive battery terminal than the shorter pin (cathode) is.
- In the schematic, the battery is shown using four lines (two short and two long). This represents a multi-cell battery, which is what a 9 V battery is. If we were using a 1.5 V battery, we would only show one long line and one short. This is because a 1.5 V battery is a single cell battery. In fact, a 9 V battery is constructed using six 1.5 V cells.

Building the Circuit on a Breadboard

By referring to either of the circuit diagrams shown above in *Figure 53*, we can build the circuit on a breadboard.

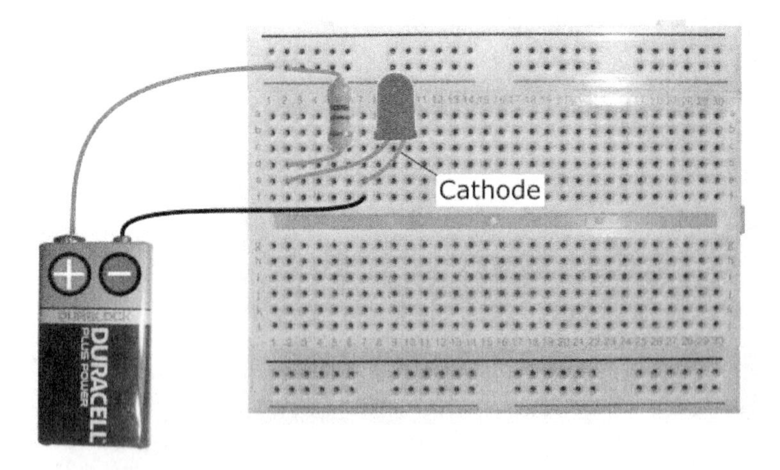

Figure 54 - The LED circuit built on a breadboard

What Happens

When electrical energy from the battery is supplied to the circuit, current flows from the positive battery terminal, through the resistor, then through the LED before making its way back to the negative battery terminal to complete the circuit.

While electric current is passing through the LED, it shines brightly. The resistor reduces the amount of current down to a suitable level to illuminate the LED, but to not damage it.

Things to Try

Try using a larger resistor than the 470 Ω one currently being used. For example, if you use a 10 kΩ resistor, the LED will still be illuminated, but much dimmer than with the 470 Ω.

Try switching the LED round so that the cathode (the shorter pin) is next to the resistor. Doing this will prevent the LED from being illuminated because the LED only lets current flow in one direction.

Try using a smaller battery supply, say, 3 V. The LED will still be illuminated, but it will be dimmer. This is because the 470 Ω resistor is way too big for a 3 V supply. Using Ohm's Law we can calculate that for a 3 V supply we only need to use a resistor of about 50 Ω.

YouTube Video

There is a YouTube video to accompany this chapter. You can find it here:

https://www.youtube.com/watch?v=yQ2-yVXFMeE&t=61s

Chapter 14

Using a Multimeter to Measure Voltage, Current and Resistance

The purpose of this chapter is to show how a digital multimeter can be used to measure resistance, voltage, and current in a breadboard circuit.

Parts Needed

We'll use the same single LED circuit that we built in the previous chapter, along with a digital multimeter.

Multimeters

You don't need to design and build electronics circuits for very long before you realize the usefulness of a multimeter. Being able to check the voltage, current, and/or resistance in different parts of a circuit will help you to understand how the circuit is working and will assist you in problem solving if there are any issues with the circuit.

What is a Multimeter?

A multimeter is a piece of equipment that combines three measuring devices:

- a voltmeter, which measures voltage,
- an ammeter, which measures current, and
- an ohmmeter, which measures resistance.

There are basically three types of multimeter to choose from:

- an analog multimeter
- a digital multimeter, and
- an auto-ranging digital multimeter.

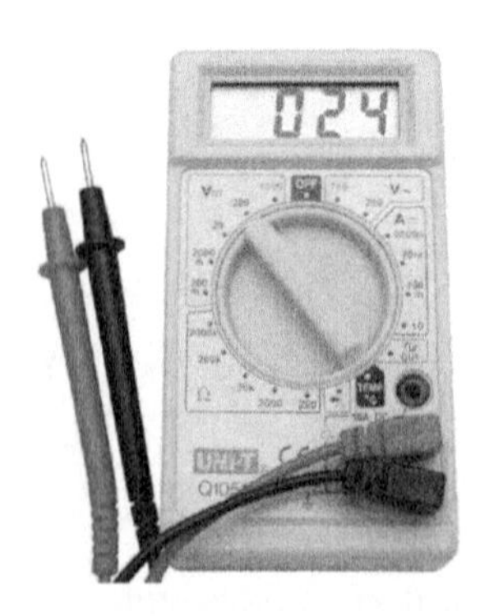

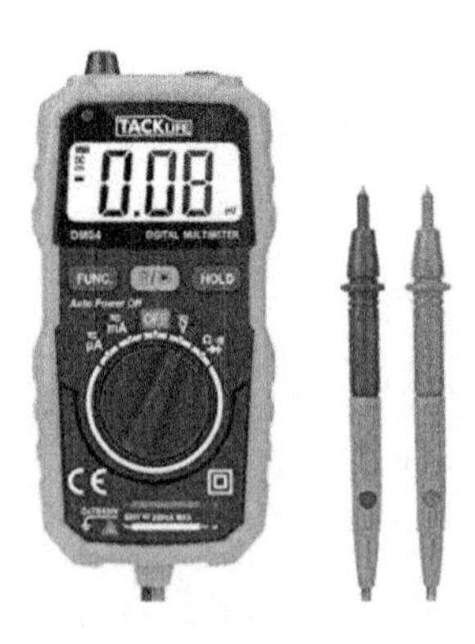

Analog Multimeter

Digital Multimeter

Auto-ranging Digital Multimeter

Figure 55 – Types of multimeter

Which Type of Multimeter Should You Buy?

Deciding which type of multimeter to use is very much a personal choice and you may well find that you start off with one type but then switch to a different one later on. I have all three types (lucky me!), but I have to admit that I never use my analog multimeter. If you're unsure what to buy, my suggestion would be to go for an auto-ranging multi-meter and see how you get on with that. You can pick them up for not very much money, so it's no great financial loss if you then decide you buy a better one at a later date.

In this chapter we'll examine how to use both an auto-ranging and non-auto-ranging digital multimeter, but not an analog one.

Auto-Ranging and Non-Auto-Ranging Digital Multimeters

With a normal (non) auto-ranging multimeter you need to set the meter to a specific range. For example, when measuring resistance, you need to set it so that it can measure up to 200 Ω for example, or up to 2 kΩ, up to 20 kΩ, up to 200 kΩ, and so on. So, if you're not sure what the resistance is that you going to measure, you would initially set the meter to a high resistance and then take a reading. If necessary, you could then change it to a lower range and take another reading. You can continue doing this until you get a sensible reading on the screen, in other words, a reading that doesn't look like this: 0.001, but more like this: 1.0.

With an auto-ranging multimeter on the other hand, you just have to set it to resistance and leave it to the meter to display the measurement in a sensible form.

Note

Your multimeter will come with its own user documentation. Make sure you read the safety information and instructions contained in that documentation. The instructions contained in this chapter relate specifically to the simple LED circuit we built in the previous chapter, and are designed to compliment rather than replace those that came with your device.

I have tried to make the instructions as meaningful as possible, but as I don't know which multimeter you are using, it may be that some information I provide does not apply to your device.

Measuring Resistance

As mentioned above, we'll use the same circuit as we did in the previous chapter. Here it is:

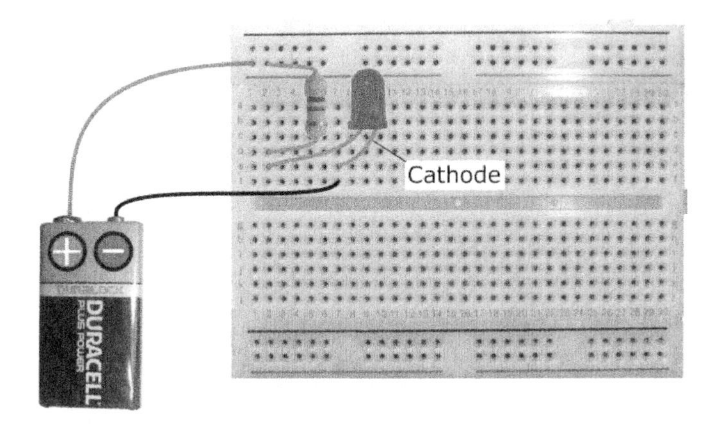

Figure 56 - A simple resistor and LED circuit

Measuring the Resistance value of the Resistor

To measure the resistance value of the resistor:

1. Disconnect the battery from the circuit. The multimeter itself will provide the necessary voltage in order to measure the resistance.

2. For a non-auto-ranging meter, connect the black test lead to the COM jack and the red (+) test lead to the V/Ω jack. For an auto-ranging meter, the leads are permanently connected so you don't need to do this.

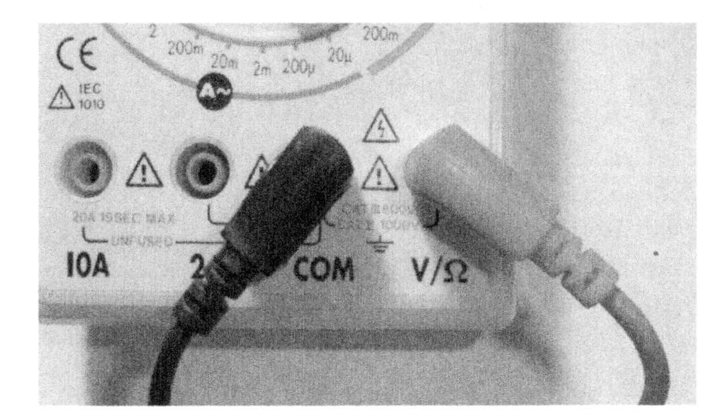

Figure 57 - Connecting the leads - voltage/resistance setting

3. Set the rotary switch to the desired Ω position. From the previous chapter we know that the resistor's value should be about 470 Ω, so set the switch to a position that is higher than that, for example, 2K as shown here:

Figure 58 - Setting the rotary switch to a suitable resistance range

For an auto-ranging multimeter you just need to set the rotary switch to Ω - you don't need to worry about selecting a range.

4. Switch on the multimeter. (*It may be that the meter already switched itself on in step 3 above.*)

5. Put one probe on one side of the resistor, and the other probe on the other side of the resistor.

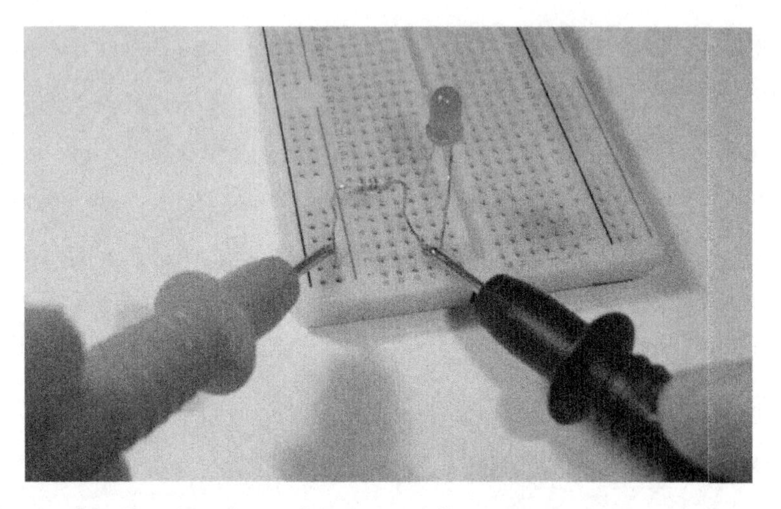

Figure 59 - Putting the probes across the resistor

It doesn't matter which probe you use on which resistor pin.

6. Read the resistance value on the multimeter's display.

Figure 60 - The resistance value

So you can see that the true value of the resistor is 466 Ω, which isn't far off the advertised value of 470 Ω.

Measuring Voltage

To measure the voltage between two points in the circuit:

1. Make sure that power is being supplied to the circuit. In other words, make sure the battery is connected and switched on.

2. For a non-auto-ranging meter, connect the black test lead to the COM jack and the red (+) test lead to the V/Ω jack. For an auto-ranging meter, the leads are permanently connected so you don't need to do this.

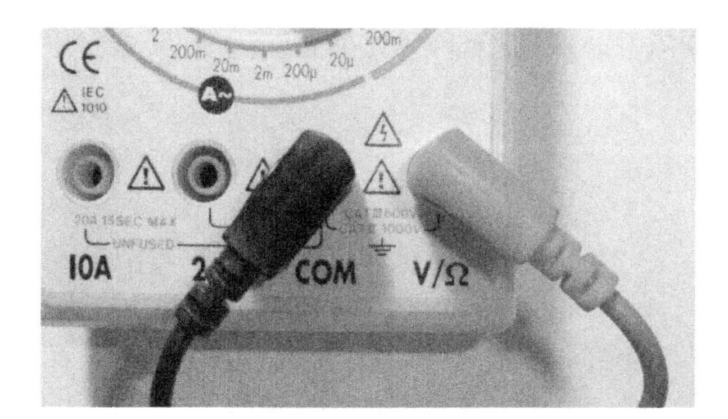

Figure 61 - Connecting the leads - voltage/resistance setting

3. Set the rotary switch to the desired position. As the voltage supply is a 9V battery, the maximum voltage across any two points in the circuit should in theory be no more than this value. However, you might find that the voltage across the battery's terminals is slightly more than 9 V - it may even be just over 10 V. So, initially at least, set the rotary switch to a voltage range that includes voltages above 10 V.

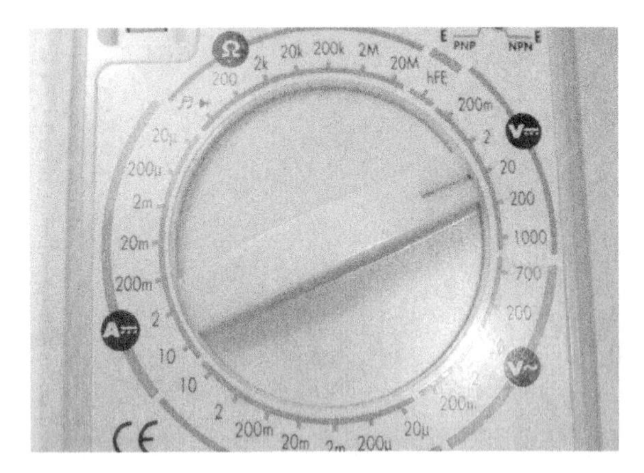

Figure 62 - Setting the rotary switch to a suitable voltage range

> DC not AC
>
> *Make sure you select DC voltage and not AC voltage.*

With an auto-ranging multimeter, you just need to select *voltage* on the rotary switch. The device will be able to cope with any voltage values (within the specifications of the device) and should be able to automatically deal with either DC or AC.

4. Switch on the multimeter. (*It may be that the meter already switched itself on in step 3 above.*)
5. Touch the probes on different points in the circuit to measure the voltage. For example, across the LED, across the resistor, or across the whole circuit (touch one probe on the positive battery terminal and the other probe the negative battery terminal).

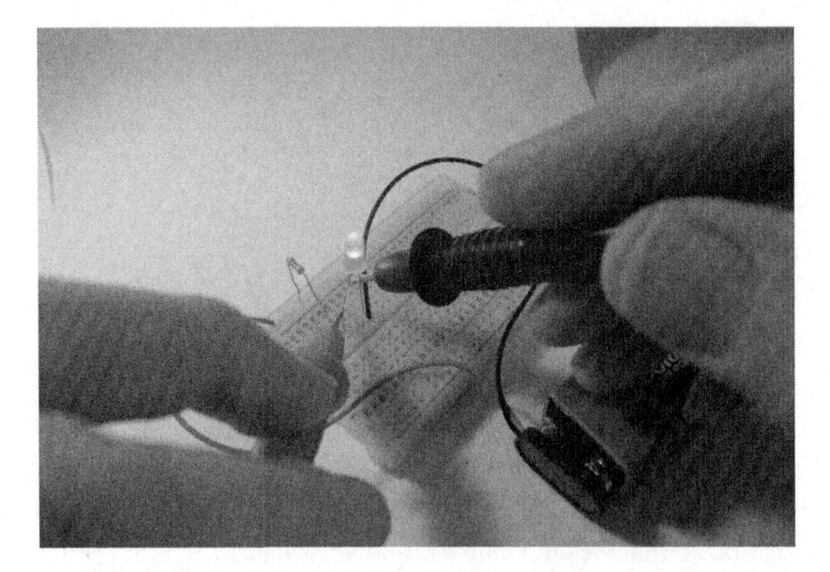

Figure 63 - Measuring the voltage across the LED

Note

Voltage is the difference in electric charge between two points, so it doesn't matter which way round you use the probes.

6. Read the voltage value on the multimeter's display. Your voltage readings should be approximately:

   - Across the whole circuit: about 9 V.
   - Across the LED: about 2 V
   - Across the resistor: about 7 V

Measuring Current

To measure the current flowing through part of a circuit, the multimeter needs to become part of the circuit. For example, if you want to know how much current is flowing through the LED, you need to remove the LED from the circuit and put the multimeter in its place, with one probe where the anode pin was connected to the breadboard, and the other probe where the cathode pin was connected.

Blowing a Fuse

When you are measuring the current flowing through a circuit, it is possible to blow a fuse in the multimeter. For example, if your multimeter is rated at 2 A / 250 V, and you place the probes across the battery terminals, you will most likely blow the multimeter's fuse because the short-circuit current will probably be several amps ... maybe as much as 8 or 9 A ... much more than the 2 A that the multimeter can cope with.

I've done this myself. It's no big deal really and you can pick up replacement fuses for next to nothing. It's more embarrassing than anything else.

To measure the current flowing between two points in the circuit:

1. Make sure that power is being supplied to the circuit. In other words, make sure the battery is connected and switched on.

2. For a non-auto-ranging meter, connect the black test lead to the COM jack and the red (+) test lead to a suitable current jack, say 2 A. For an auto-ranging meter, the leads are permanently connected so you don't need to do this.

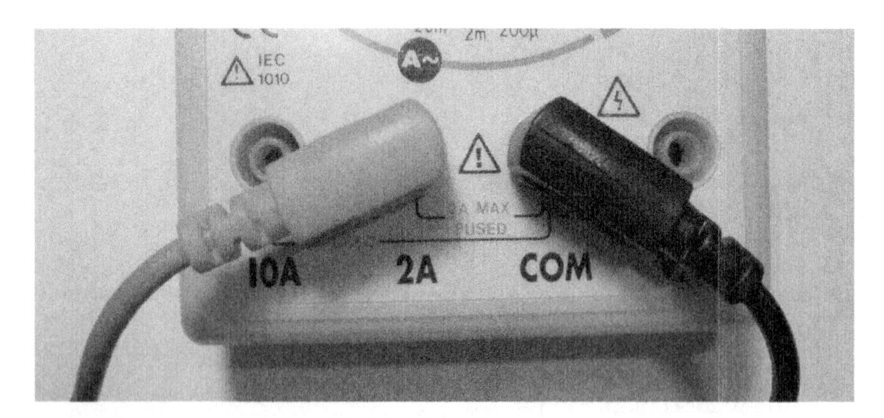

Figure 64 - Connecting the leads - 2 A current setting

3. Set the rotary switch to the desired position - something in the region of 20 or 200 mA should suffice.

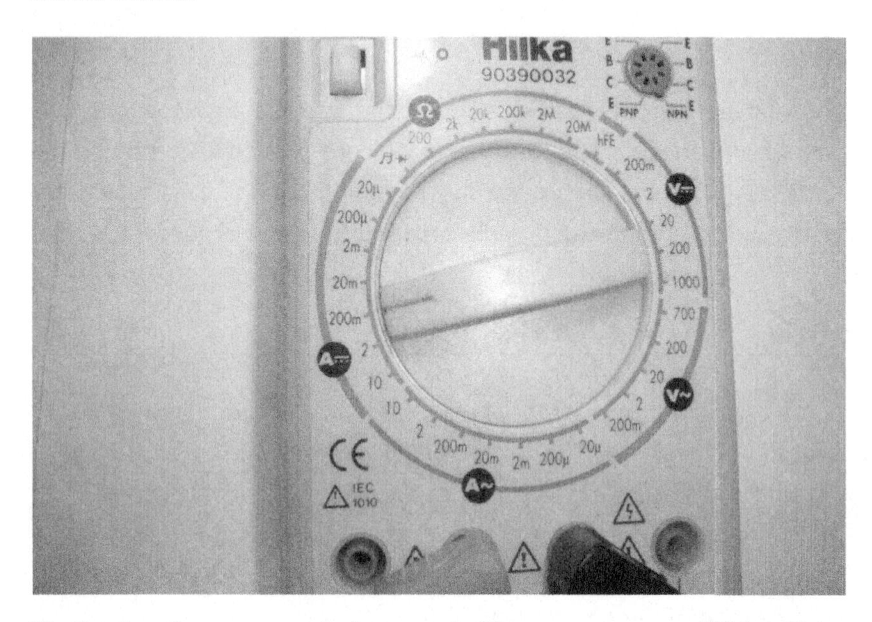

Figure 65 - Setting the rotary switch to a suitable current range (200 mA)

> *DC not AC*
>
> *Make sure you select DC current and not AC current.*

If you're using an auto-ranging multimeter, you'll probably just need to select either µA or mA.

4. Switch on the multimeter. (*It may be that the meter already switched itself on in step 3 above.*)

5. Decide which part of the circuit you want to check, for example, the LED, and then remove that part of the circuit.

6. Insert the probes of the multimeter into the circuit, in place of the part you removed in Step 5.

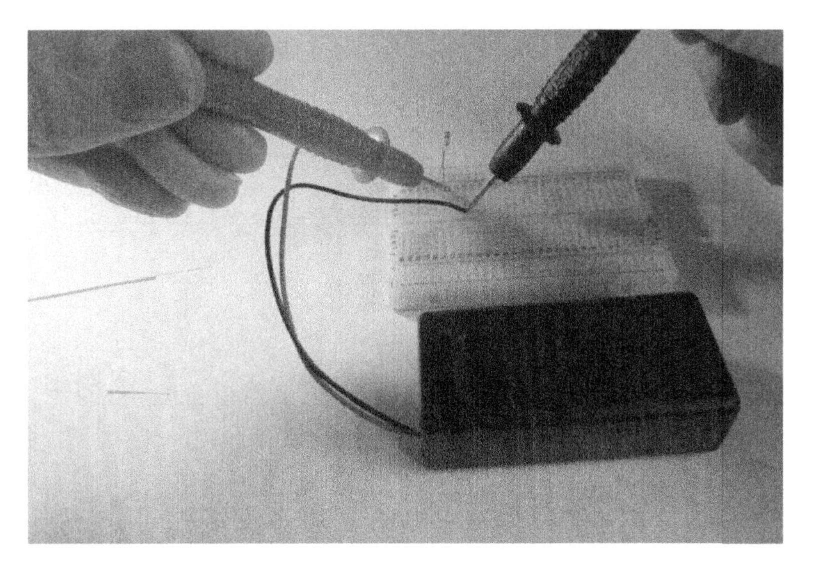

Figure 66 - Measuring the current that flows through the LED

> *Note*
>
> *It doesn't matter which way round you insert the probes.*

7. Read the current value on the multimeter's display. For a 9 V battery and 470 Ω resistor, the current will be somewhere in the region of 18-20 mA.

YouTube Videos

There are a few YouTube videos to accompany this chapter. You can find them here:

Measuring Resistance

https://www.youtube.com/watch?v=quW1dKtpvG4

Measuring Voltage

https://www.youtube.com/watch?v=wZRF1ZStVfk

Measuring Current

https://www.youtube.com/watch?v=eymN9PJ7sFw

Chapter 15

Connecting Multiple LEDs in Series

The purpose of this experiment is to extend the circuit we designed and built in Chapter 13 on Page 93 by examining how to add multiple LEDs in series.

Parts Needed

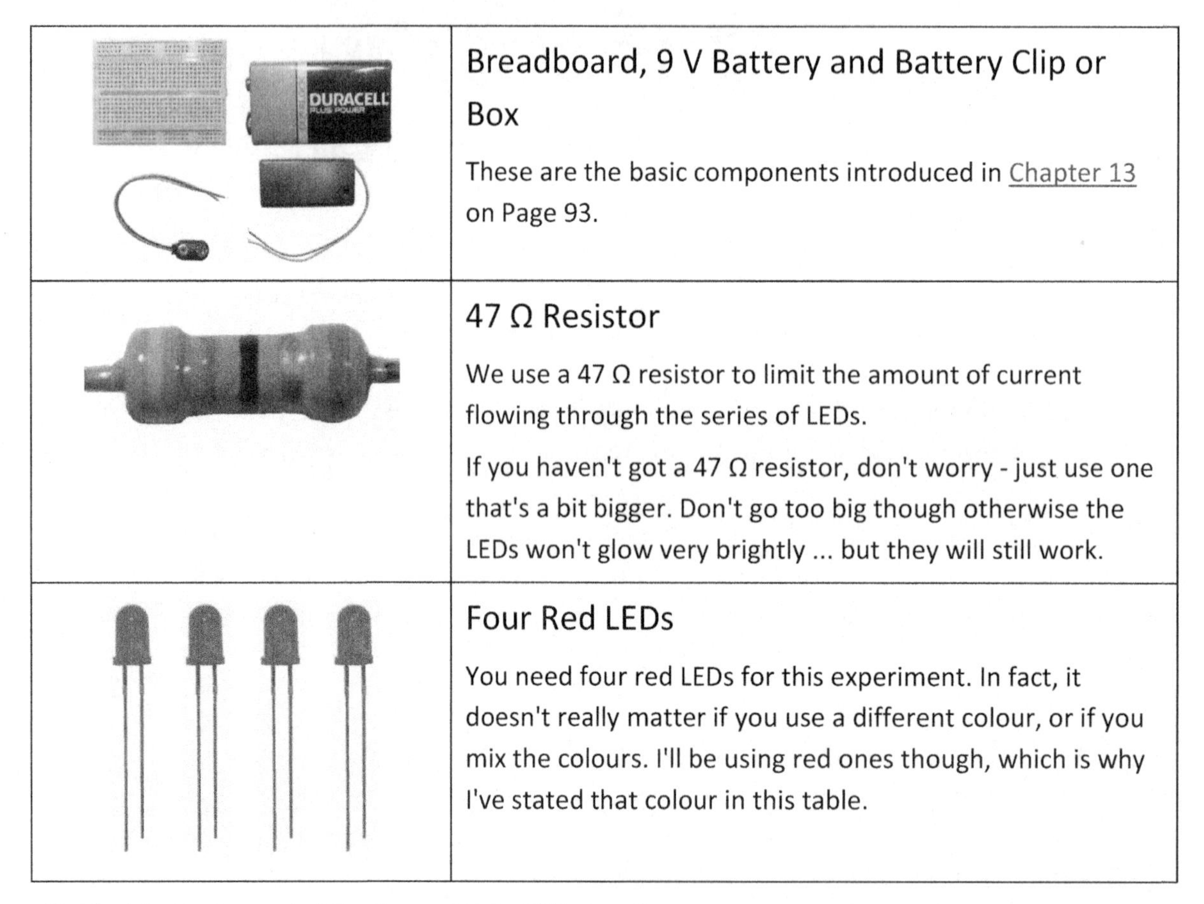

	Breadboard, 9 V Battery and Battery Clip or Box These are the basic components introduced in Chapter 13 on Page 93.
	47 Ω Resistor We use a 47 Ω resistor to limit the amount of current flowing through the series of LEDs. If you haven't got a 47 Ω resistor, don't worry - just use one that's a bit bigger. Don't go too big though otherwise the LEDs won't glow very brightly ... but they will still work.
	Four Red LEDs You need four red LEDs for this experiment. In fact, it doesn't really matter if you use a different colour, or if you mix the colours. I'll be using red ones though, which is why I've stated that colour in this table.

Table 18 - Parts needed for the multiple LED circuit

Using Ohm's Law to Calculate the Correct Resistor to Use

In Chapter 13 on Page 93 we spent quite a bit of time looking at how Ohms Law can be used to calculate the size of the resistor needed to protect the LED, which turned out to be 470 Ω. In this chapter, we will again use Ohm's Law to calculate the resistor size.

Here is Ohm's Law:

Resistance = Voltage / Current

When LEDs are connected in series, their individual voltage drops are added together to give an overall voltage drop. If the voltage drop for a single LED is 2 V then the voltage drop for four LEDs is 8 V.

Here is the slightly modified formula to use:

Resistor Value = [Power Supply Voltage - (LED voltage drop x number of LEDs)] / Current

So, assuming a current of 20 mA (0.020 A), we have:

Resistor Value = [9 - (2 x 4)] / 0.020

Resistor Value = 50 Ω

As I don't have a 50 Ω resistor, I'll be using a 47 Ω one, which I do have.

Designing the Circuit

Let's start by taking a quick look back at the circuit we built in Chapter 13 on Page 93 in which a single LED is illuminated when a current passes through the circuit. For convenience, I've reproduced it below.

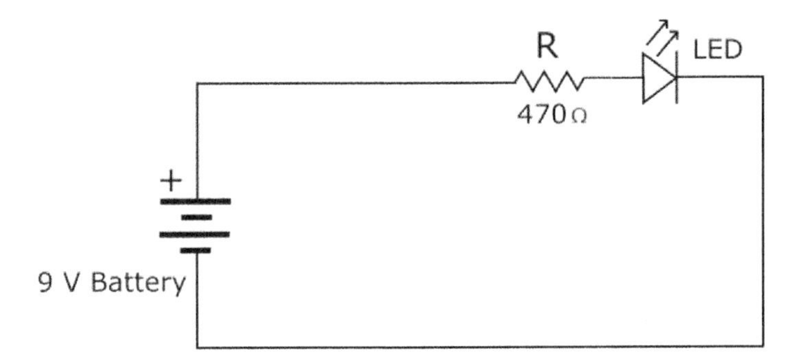

Figure 67 - A circuit to illuminate a single LED

What we are going to do in this chapter is to extend this circuit so that multiple LEDs (positioned in series) are illuminated when power is supplied to the circuit. The modified circuit is shown below in *Figure 68*.

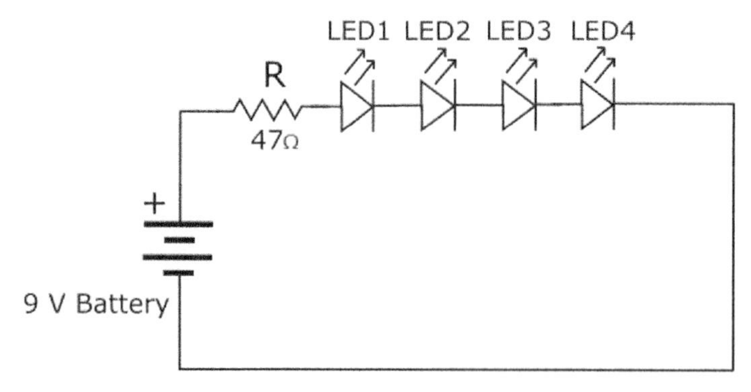

Figure 68 - Multiple LEDs in series circuit

Building the Circuit on a Breadboard

Now that we know what the circuit looks like, we can build it on a breadboard, as shown below in *Figure 69*.

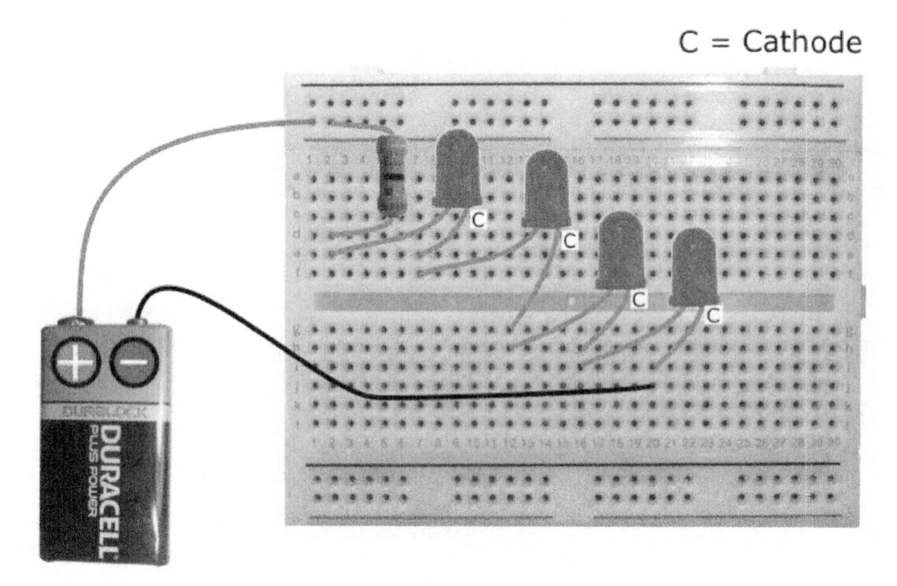

Figure 69 – Multiple LEDs in series circuit

The circuit is very easy to build on a breadboard. Just be sure to insert the LEDs the right way round otherwise the circuit won't work.

What Happens

When electrical energy from the battery is supplied to the circuit, current flows from the positive battery terminal, through the resistor, then through the four LEDs before making its way back to the negative battery terminal to complete the circuit.

Things to Try

Try adding a fifth LED. If you do this you no longer need to use a resistor to protect the LEDs because the voltage drop across them is sufficient to match the supply voltage. In fact, the battery will probably struggle to illuminate the LEDs brightly. If you add a sixth LED, a 9 V supply almost certainly won't be up to the job. When I tried six LEDs, two of them were just about illuminated when viewed in the dark. I guess these two were from a different batch and required slightly less current to operate.

Voltage Drop Change

The voltage drop across an entire DC circuit equals the supply voltage (Kirchhoff's circuit laws). Therefore, if the supply voltage is 9 V, the total voltage drop of all the components (resistors, LEDs, jumper wires, and so on) will also equal 9 V. Therefore, as you add more LEDs to a circuit, their individual voltage drops are reduced.

Example:

The following voltage drop values are from measurements I took myself by building a circuit on a breadboard that powered one LED, and then gradually adding more LEDs.

Assuming a 470 Ω resistor and 9 V supply voltage:

- *For one LED in the circuit, the voltage drop across the LED is 1.98 V*
- *For two LEDs, the voltage drop across each LED is 1.95 V*
- *For Three LEDs, the voltage drop across each LED is 1.90 V*
- *For four LEDs, the voltage drop across each LED is 1.85 V*
- *For five LEDs, the voltage drop across each LED is 1.74 V*
- *For six LEDs, the voltage drop across each LED is 1.50 V*

Of course, with different LEDs, your results may well be different, but you can see that the voltage drop is gradually reducing.

YouTube Video

There is a YouTube video to accompany this chapter. You can find it here:

https://youtu.be/O9Mh1tM9Wr0

Chapter 16

Connecting Multiple LEDs in Parallel

The purpose of this experiment is to extend the circuit we designed and built in Chapter 13 on Page 93 by examining how to add multiple LEDs in parallel.

Parts Needed

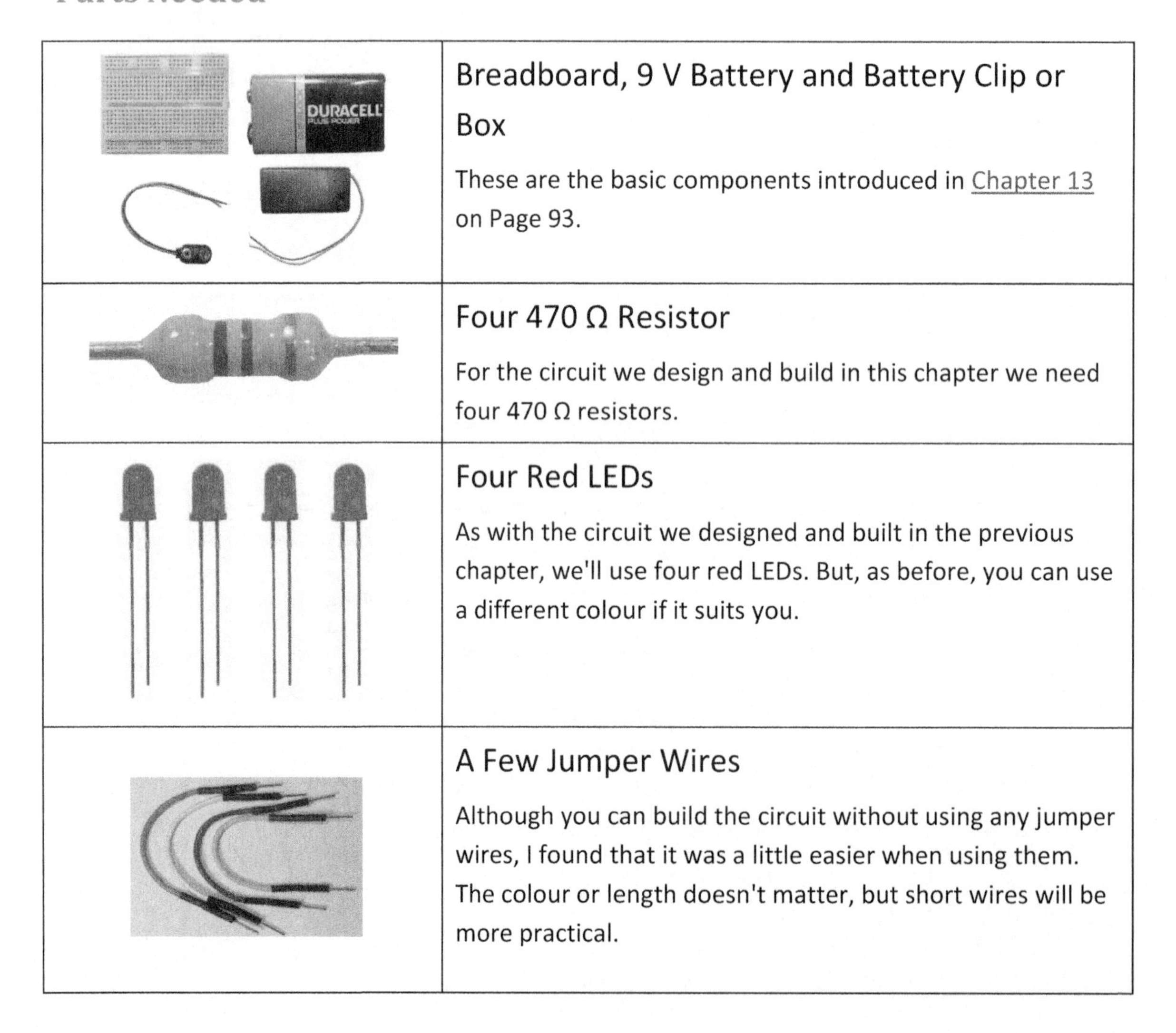

	Breadboard, 9 V Battery and Battery Clip or Box These are the basic components introduced in Chapter 13 on Page 93.
	Four 470 Ω Resistor For the circuit we design and build in this chapter we need four 470 Ω resistors.
	Four Red LEDs As with the circuit we designed and built in the previous chapter, we'll use four red LEDs. But, as before, you can use a different colour if it suits you.
	A Few Jumper Wires Although you can build the circuit without using any jumper wires, I found that it was a little easier when using them. The colour or length doesn't matter, but short wires will be more practical.

Table 19 - Parts needed for the multiple LEDs in parallel circuit

Using Ohm's Law to Calculate the Correct Resistor to Use

Calculating the size of the resistors to use when connecting LEDs in parallel is the same as when a single LED is used. If you can't remember how we calculated that a 470 Ω resistor is the correct one to use then go back and have a look at the calculations we performed in Chapter 13 on Page 93.

Designing the Circuit

Once again, our starting point is the single LED circuit we designed in Chapter 13 on Page 93, shown below.

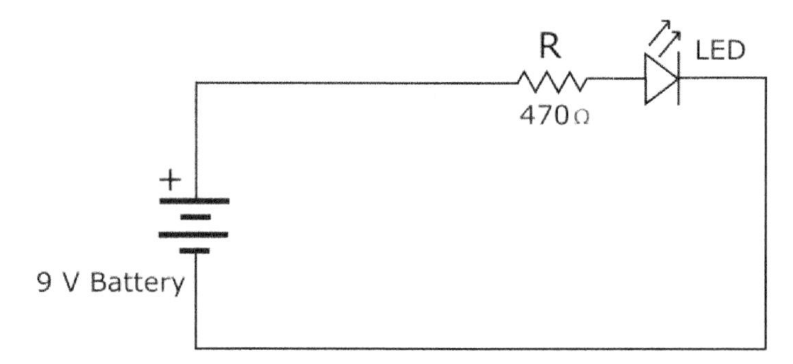

Figure 70 - A circuit to illuminate a single LED

The modified circuit for multiple LEDs in parallel is shown below in *Figure 71*.

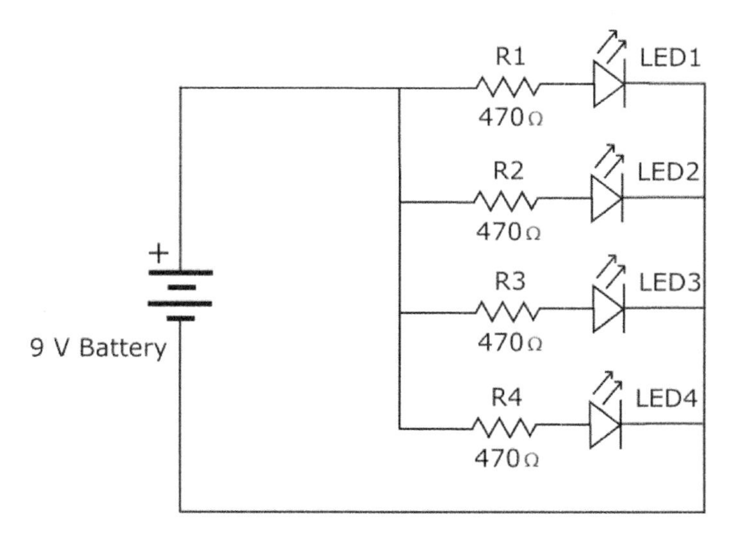

Figure 71 - Multiple LEDs in parallel circuit

Alternative Circuit

Although you can, you should probably avoid connecting multiple LEDs in parallel with a single resistor shared between them, especially if the LEDs are not identical (all red, all green, and so on). Even if they are all identical in theory, it is quite possible that the individual LEDs have slightly different specifications. The problem is that if one of the LEDs draws more current than it can cope with, it may well become damaged and stop working. This will then cause one or more of the remaining LEDs to draw more current than they did before, which again may well lead to further damage in the circuit.

However, for completeness, here is such a circuit:

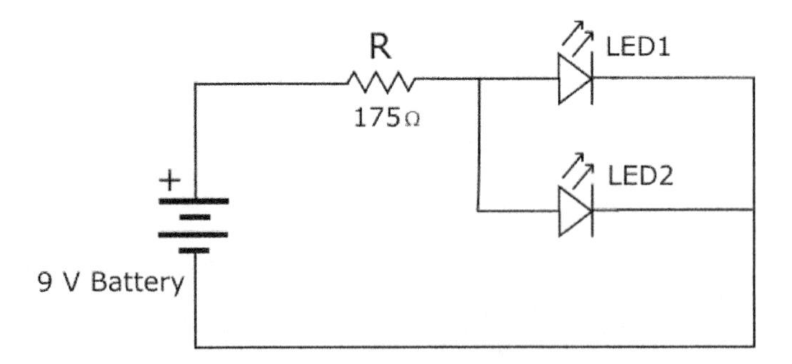

Figure 72 - Alternative multiple LEDs in parallel circuit

Assuming an LED voltage drop of 2 V, and a required current of 0.020 A for each LED, this is how we use Ohm's Law to calculate the resistor size for the above circuit:

Resistor Size = (Supply Voltage - Voltage Drop for 1 LED) / (Current x Number of LEDs)

Resistor Size = (9 V - 2 V) / (0.020 A x 2)

Resistor Size = 7 V / 0.040 A

Resistor Size = 175 Ω

Building the Circuit on a Breadboard

Now that we know what the circuit looks like (*Figure 71* on the previous page), we can build it on a breadboard, as shown below in *Figure 73*.

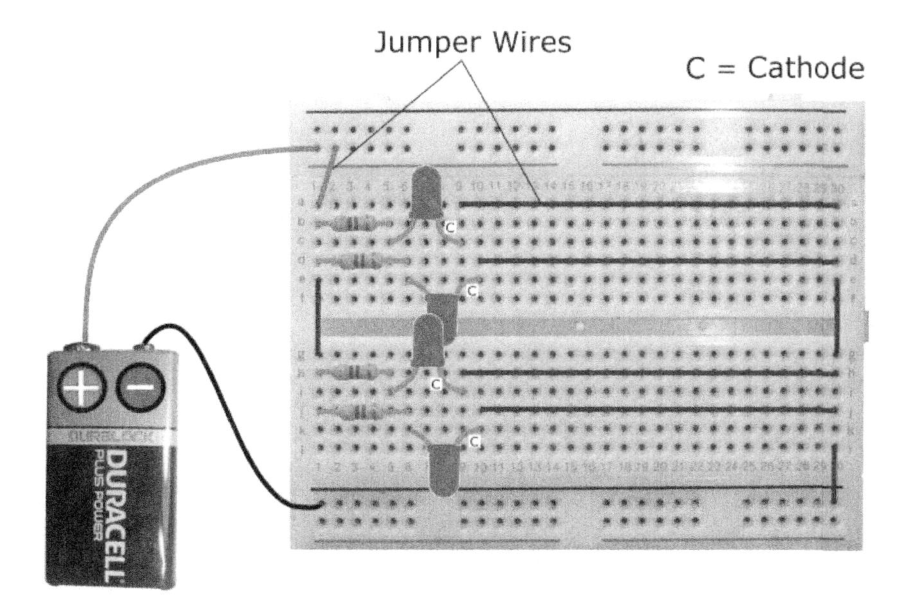

Figure 73 – Multiple LEDs in parallel circuit

The circuit is very easy to build on a breadboard. As with other LED circuits we have built, just be sure to insert the LEDs the right way round otherwise the circuit won't work.

What Happens

When electrical energy from the battery is supplied to the circuit, current flows from the positive battery terminal through each resistor and LED before making its way back to the negative battery terminal to complete the circuit.

YouTube Video

There is a YouTube video to accompany this chapter. You can find it here:

https://youtu.be/Bhv-Tk7l1zI

Chapter 17

Using a Variable Resistor to Control the Speed of a Small DC Motor

The purpose of this experiment is to show how a potentiometer can be used to control the speed of a small DC motor.

Parts Needed

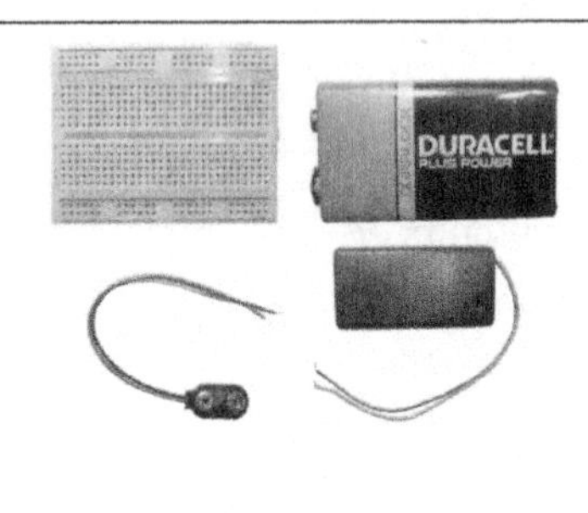	**Breadboard, 9 V Battery (see note below) and Battery Clip or Box** These are the basic components introduced in Chapter 13 on Page 93. *Note: Some very small DC motors only require a supply voltage of between 1.5 and 6 V. If you are using one of these - or think that you might be - use a smaller voltage supply - say 3 or 6 V.*
	**Potentiometer** We'll use a potentiometer wired up as a variable resistor - refer to Using a Potentiometer as a Variable Resistor (Rheostat) if you've forgotten what this means - to control the motor speed. If you can, use a potentiometer that has a fairly small value, say 1 kΩ or 5 kΩ, rather than one with a value such as 100 kΩ. This will enable you to more delicately control the speed of the motor. With a large-valued potentiometer you'll probably find that the motor is either on or off, with very little control in between.
	**2-way Terminal Block** To make it easy to connect the positive and negative wires that come from the motor to the breadboard, we'll make use of a small terminal block.

	**DC Motor and Fan Blade** DC motors come in a range of sizes. If you can, try to get a small 'hobby' DC motor that can cope with a supply voltage of 9 V (see the note above). Although not strictly necessary, it makes the experiment more visually satisfying if you can attach a small fan blade to the motor's spindle so that you can see how fast or slow the shaft is spinning.
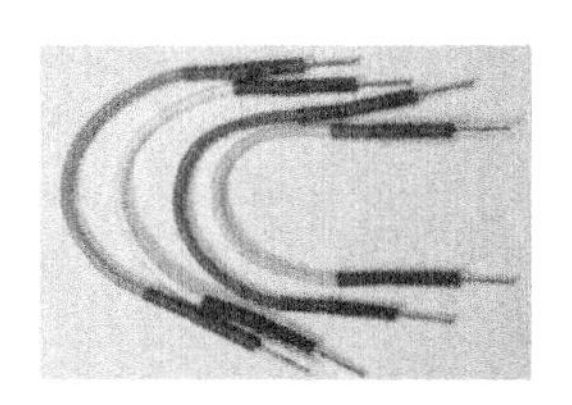	**A Few Jumper Wires** We also need a few jumper wires. The colour or length doesn't matter, but short wires will be more practical.

Table 20 - Parts needed for the DC motor circuit

Introduction to DC Motors

A DC motor is any of a class of rotary electrical machine that converts direct current (DC) electrical energy into mechanical energy. The most common types rely on the forces produced by magnetic fields.

Basic Principle of How a DC Motor Works

When an electric current passes through a wire (wound into a coil) inside a magnetic field, the magnetic force produces a torque (turning force) which turns the DC motor. A visual representation of a very basic DC motor is shown in *Figure 74*, below.

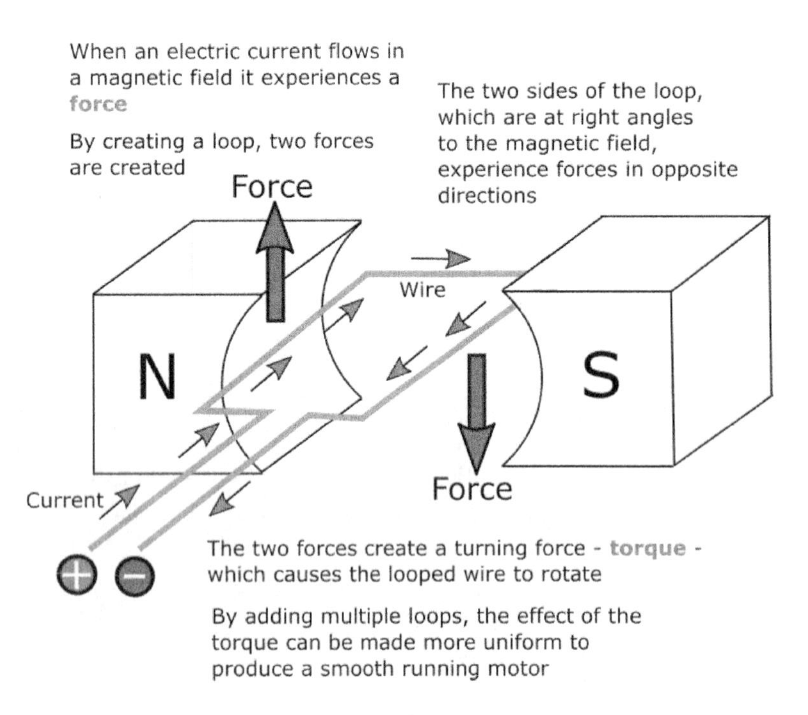

Figure 74 - How a DC motor works

Hobby DC Motors

DC Motors are available in many sizes, ranging from small *hobby* motors that run on a voltage supply as little as 1.5 V, all the way up to large motors that require 24 V or more.

As well as the voltage rating of a DC motor, there is also a current rating. For the circuit we'll be building in this chapter, we don't need to worry too much about how much current the motor is drawing because we won't be running the motor for a long period of time or applying much of a load to it (the fan blade), which increases the amount of current drawn. However, in a real life situation it is important to not let the motor draw too much current otherwise this could cause the motor to overheat or even burn out.

Table 21 shows some specifications for a small DC motor.

Supply voltage	DC 6 V - 15 V
Speed at max efficiency	8768 rpm @ 12 V
Current at max efficiency	0.758 A @ 12 V
Output power at max efficiency	58 W @ 12 V
Torque (turning force)	64.2 g.cm (0.0063 Nm)
Shaft diameter	2.3 mm
Weight	51 g

Dimensions	27.63 mm x 27.63 mm x 38.8 mm

Table 21 - Specifications for the MFA 457-RE385 DC motor

Although not shown in the above table, motors often have two current ratings: one for when the motor is not driving a load, and one for when it is (sometimes called the *stall* load).

It is important to remember that the current drawn when the motor has no load on it is less than the current drawn when the motor is under load or in a stalled (no longer turning) condition.

If a DC motor has a supply voltage range of 6 to 15 V, and you run it at the lower end (6 V), the amount of current it requires to operate at this voltage is more than if you use a higher supply voltage. It is therefore better (more efficient) to run it at a higher voltage, say, 12 V as is shown in the example motor specifications shown in *Table 21*.

Controlling the Direction of a DC Motor

To change the direction in which the shaft rotates you can switch the polarity of the power connections to the motor. To do this you can simply swap the positive and negative wires over. There are more sophisticated ways to do this, for example, by using a double-pole, double-throw (DPDT) switch.

Controlling the Speed of a DC Motor

There are various ways to control the speed of a DC motor, for example, you can use a set of gears, you can use Pulse Width Modulation (PWM), and you can use a variable resistor connected in series with the motor, which is the method we will look at in this chapter.

Controlling the Speed by Using a Variable Resistor (Rheostat)

You can control the speed of a DC motor by varying the amount of current supplied to it. Although this can be done very easily using a variable resistor, which is what we'll be doing in this chapter, in a real-life situation this would probably not be the best way to control the motor speed.

One problem with using a resistor is that the current drawn by the motor increases as the load on the motor increases. More current means a larger voltage drop across the resistor and therefore even less voltage to drive the motor. This causes the motor to draw even more current, possibly resulting in the motor stalling.

Designing the Circuit

Anyway, ignoring the drawbacks of using a resistor to control motor speed, we can move on to designing the circuit.

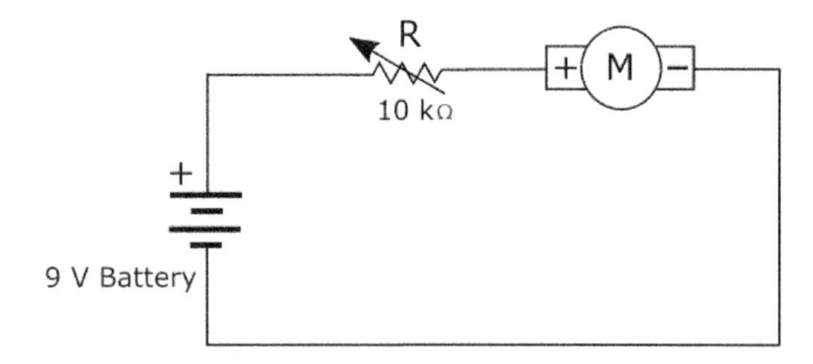

Figure 75 - Speed control of a DC motor (schematic circuit)

Building the Circuit on a Breadboard

Having designed the circuit, we can now build it on a breadboard as shown below in *Figure 76*.

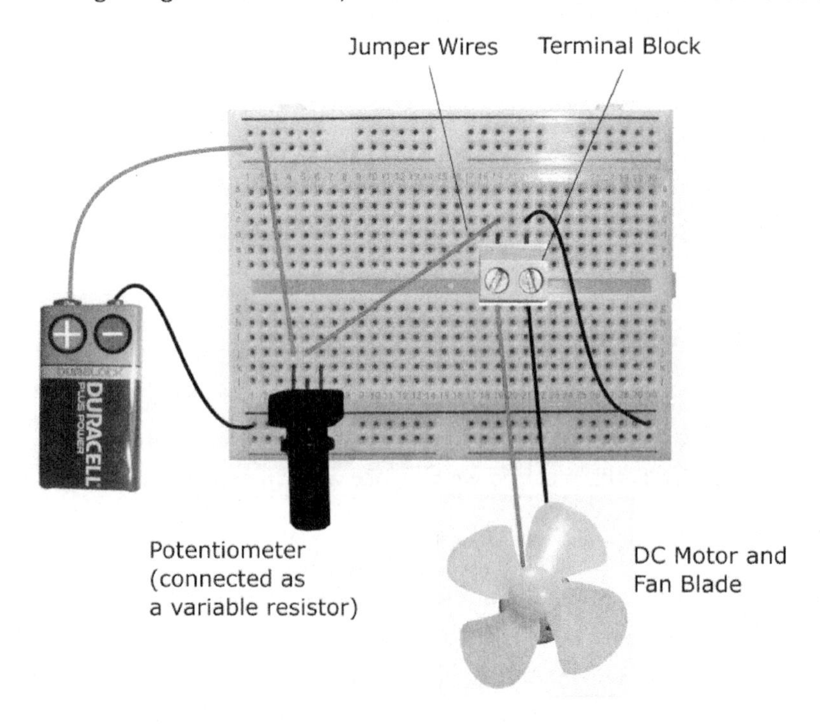

Figure 76 - Speed control of a DC motor (breadboard circuit)

Potentiometer Wired Up as a Variable Resistor

If you look at the potentiometer you can see that only two of the pins are connected - the middle one and one of the outer ones. This means it is wired up as a variable resistor. It doesn't matter which outer pin is used, it just means that the effect of turning the knob on the potentiometer is reversed if you use one pin as opposed to the other.

If all three pins were connected, the potentiometer would act as a voltage divider. In this situation, the middle pin provides the voltage and current output (as it does for a variable resistor), while one of the outer pins would be connected to the voltage supply (the battery's positive terminal) with the other being connected to the negative terminal.

What Happens

By using a variable resistor, the amount of current passing through the wire can be varied, thus causing the speed of the motor to vary.

Things to Try

You could try replacing the variable resistor with a fixed value one to fix the motor speed.

YouTube Video

There is a YouTube video to accompany this chapter. You can find it here:

https://youtu.be/VnV2WW34bg8

Chapter 18

Wind Power

The purpose of this experiment is to show how a DC motor can be used to generate an electric current.

In the previous chapter we learned that if a wire carrying an electric current is placed inside a magnetic field it produces torque (turning force) that can be used to turn a DC motor. Well, the opposite is also true - if torque is applied to the DC motor, an electric current is generated.

Parts Needed

	Breadboard As usual, you build the circuit on this.
	470 Ω Resistor As we'll be illuminating an LED, we'll need to protect it in the same way as we have in previous experiments. Exactly how much voltage and current are generated will depend to a large extent on how hard you can blow, but as these values may well be in the region of 6 V and 50 mA*, we'll use a 470 Ω resistor, which is the size we use when we're using a 9 V battery as the power supply. * Although you wouldn't want to run the LED at a continuous current of 50 mA, giving it a quick burst at this level of current is unlikely to damage it. If it does, just use another LED and don't blow so hard!
	One LED You need one LED - it doesn't matter what colour it is.

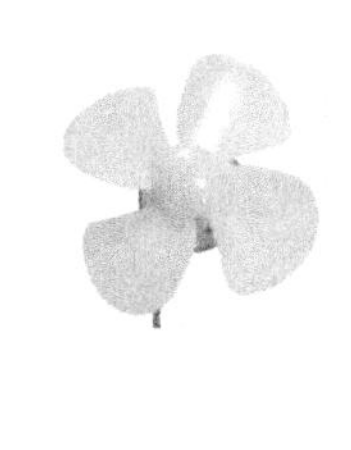	**DC Motor and Fan Blade** As discussed in the previous chapter, DC motors come in a range of sizes. For this experiment we just need a small hobby DC motor - somewhere in the range 1.5 V to 6 V. However, whereas in the previous experiment it wasn't essential that you had a fan blade, in this experiment it is - without one, there'll be nothing to blow!
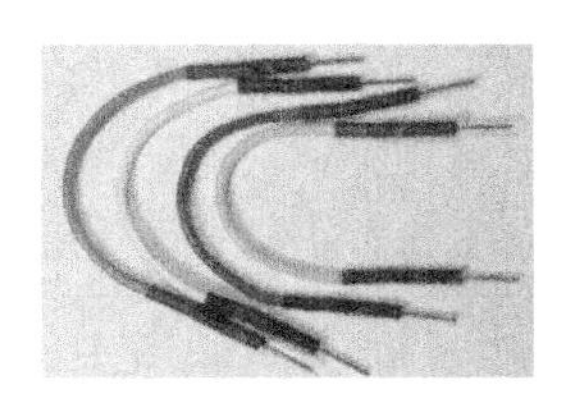	**One Jumper Wire** Although you can build this circuit without using any jumper wires, using one will make it slightly easier to arrange the components. The colour or length doesn't matter, but a short one will be more practical.

Table 22 - Parts needed for the wind power circuit

Designing the Circuit

The circuit is very simple, as shown below.

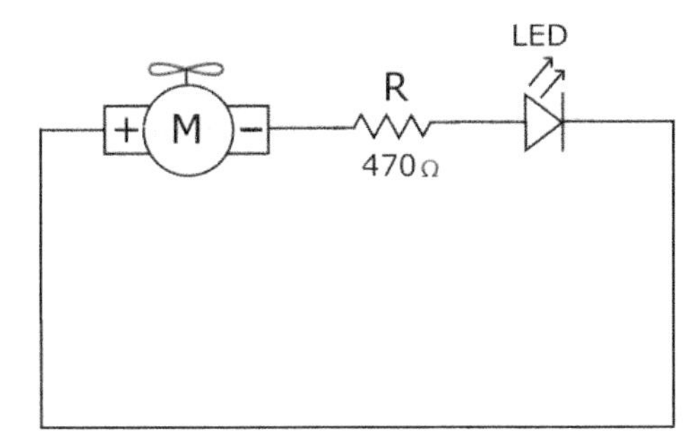

Figure 77 - Wind power (schematic circuit)

With reference to *Figures 77* (above) and *78* (below), the only thing that you may perhaps think is a bit odd with this circuit is that the resistor and LED are connected to the negative lead coming from the motor. If, when you build the circuit yourself, it doesn't work, try connecting the resistor and LED to the positive lead coming from the motor.

The polarity of a DC motor isn't fixed in the same way as it is for a battery, and you can reverse it by moving the supply voltage from one terminal to the other.

Building the Circuit on a Breadboard

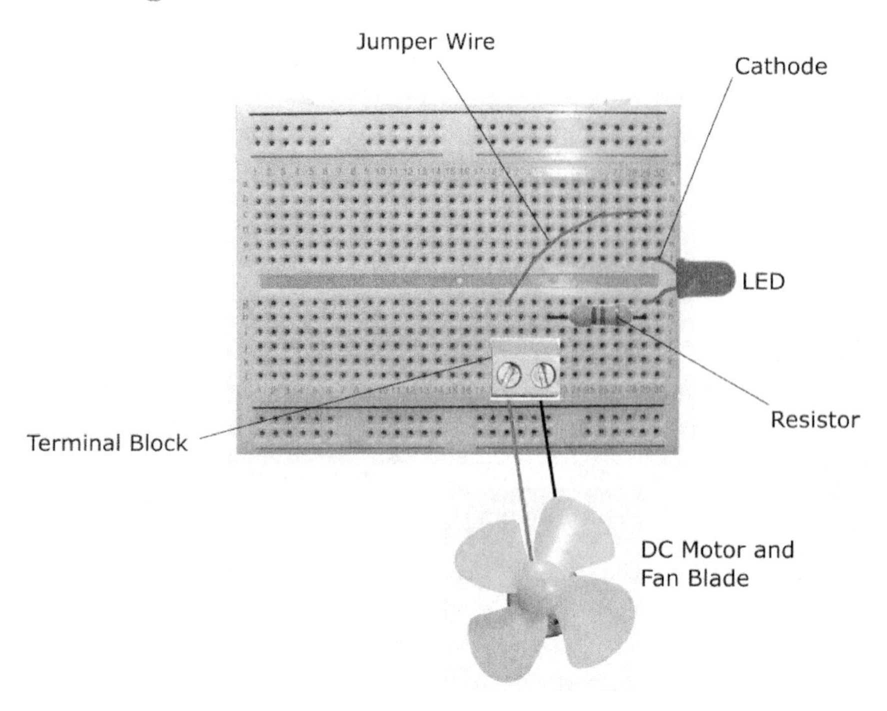

Figure 78 - Wind power (breadboard circuit)

What Happens

As you blow on the fan, it turns the motor's spindle. The torque generated by doing this causes an electric current to flow down the wire and illuminate the LED.

YouTube Video

There is a YouTube video to accompany this chapter. You can find it here:

https://youtu.be/R3Bbs3iKZzo

Chapter 19

Using a Transistor to Switch On an LED

The purpose of this experiment is to show how a transistor can be used to switch on an LED, and in so doing, demonstrate how a small current can be used to control the flow of a larger current.

Parts Needed

	Breadboard, 9 V Battery and Battery Clip or Box As for previous experiments.
	470 Ω Resistor We use a 470 Ω resistor to limit the amount of current flowing through one of the LEDs.
	14.7 kΩ Resistor This is the resistor we use to control how much current flows to the Base of the transistor. You don't have to use a 14.7 kΩ resistor if you don't have one - anything in the range 470 Ω to about 75 kΩ will probably do, but something around 10-15 kΩ is ideal.
	Push Button We use a push button to let current flow to the Base of the transistor (via the 14.7 kΩ resistor and one of the LEDs).
	NPN Transistor In the circuit described in this chapter we use a BC546 B NPN transistor, but the circuit will probably work with a wide range NPN transistors.

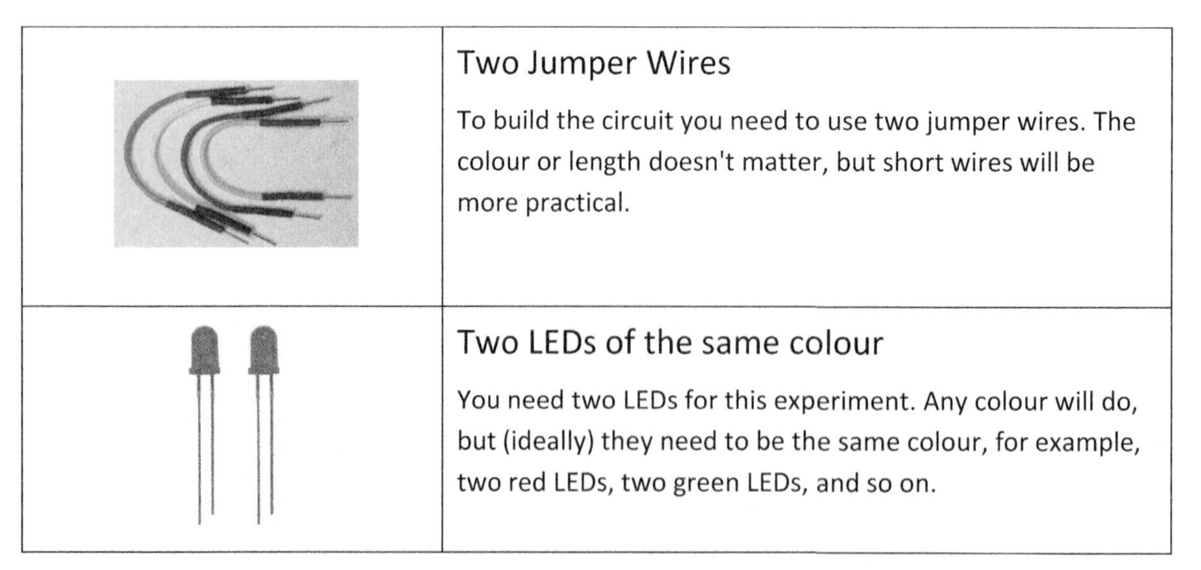

	Two Jumper Wires To build the circuit you need to use two jumper wires. The colour or length doesn't matter, but short wires will be more practical.
	Two LEDs of the same colour You need two LEDs for this experiment. Any colour will do, but (ideally) they need to be the same colour, for example, two red LEDs, two green LEDs, and so on.

Table 23 - Parts needed for the transistor-LED circuit

Designing the Circuit

The starting point for this experiment is the circuit we built in Chapter 13 on Page 93 in which a single LED is illuminated when a current passes through the circuit – repeated below for convenience.

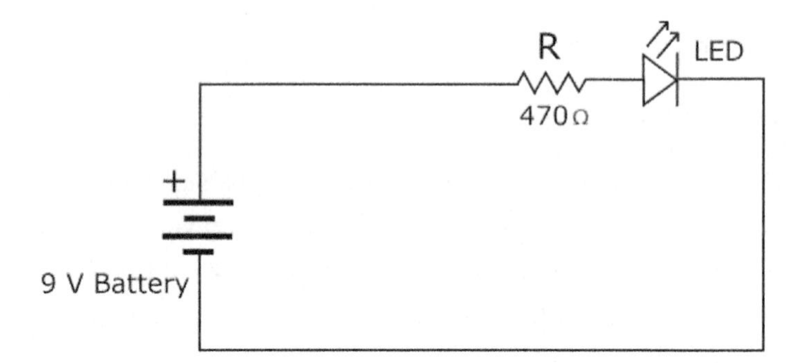

Figure 79 - A simple LED circuit

If you can't remember how the above circuit works, go back to Chapter 13 on Page 93 now to refresh your memory.

What we are going to do in this chapter is to extend this circuit so that a transistor controls whether or not the LED is illuminated. The modified circuit is shown below in *Figure 80.*

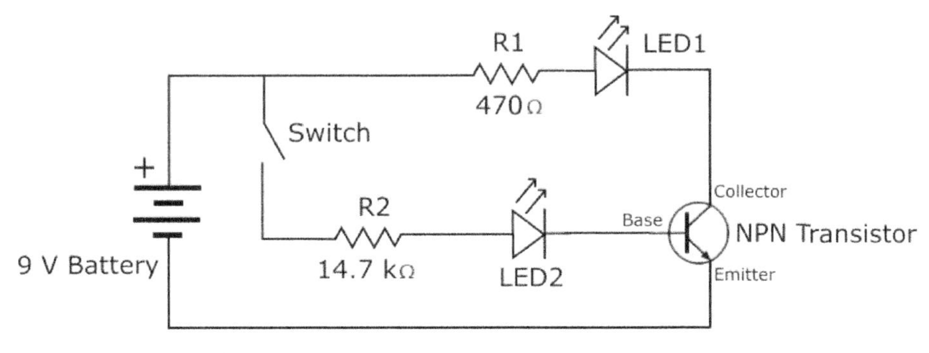

Figure 80 - Modified LED circuit

Looking at *Figure 80* you can see that the basic LED circuit has been modified to add a second resistor and LED, a switch, and an NPN transistor. The original LED (now labeled as LED1) will only illuminate when the transistor is switched on by applying power (voltage and current) to the transistor's Base.

The purpose of LED2 in the circuit is simply to demonstrate that the current being applied to the Base is very small - LED2 will only just about be illuminated when the switch is closed, whereas LED1 will be illuminated to the same intensity as it was previously (before the circuit was modified).

Calculating the Current through the Base-Emitter Path (I_{BE})

To calculate the current being applied to the Base we can use Ohm's Law:

$I_{BE} = V / R$

I_{BE} - (9 - 2) / 14700 (9 V = battery voltage, and 2 V is the forward voltage drop across LED2)

$I_{BE} = 0.0004761$ A or 0.5 mA

If you remember from Chapter 13 on Page 93, an LED typically needs about 20 mA to shine brightly. The current passing through LED2 is only 0.5 mA, which will cause it to be very dim when the circuit is connected up.

If you use a bigger resistor for R2, the I_{BE} current will be even smaller. In fact you could go much bigger with the resistor if you wanted to and the transistor would still operate.

Maximum values

Be aware with transistors that there are maximum power (in watts), voltage and current values above which the transistor will be damaged or destroyed. Refer to the data sheet supplied with your transistor(s) for more information.

Building the Circuit on a Breadboard

Having designed the circuit, we can now build it on a breadboard as shown below in *Figure 81*.

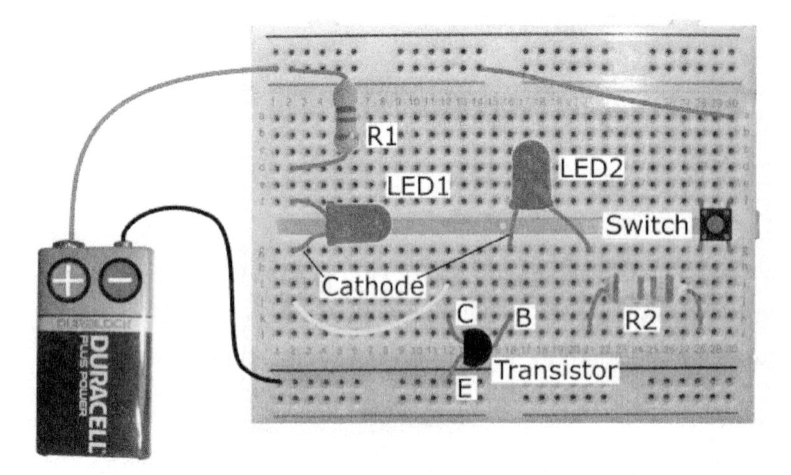

Figure 81 - Building the circuit on a breadboard

Component positioning

If you have trouble positioning some of the components exactly as shown in Figure 81, move them so that you can insert them into the breadboard. For example, you might find that the transistor pins are too short to insert them as shown, in which case you can use different holes in the breadboard. The important (essential) thing though is to make sure that you don't break the circuit.

What Happens

The purpose of the circuit is to illuminate LED1 (as previously mentioned, LED2 is included simply to show that the current flowing to the Base is very small).

In its stable state - when the push button is not pressed - current is not able to flow completely through the circuit from the battery's positive terminal to the negative terminal. There are two routes for the current to take:

1. Through R1, then LED1, along the yellow jumper wire, and then onto the transistor's Collector pin. This is where the current stops along this route.

2. Along the red jumper wire and then down to the switch. This is where the current stops along this route.

When the switch is closed (see *Figure 82*, below), current is able to flow along both routes. Firstly, current flows through the switch, then through R2, then LED2, and finally down to the

Base pin on the transistor. When this happens, the transistor lets current flow from the Collector to the Emitter, which illuminates LED1.

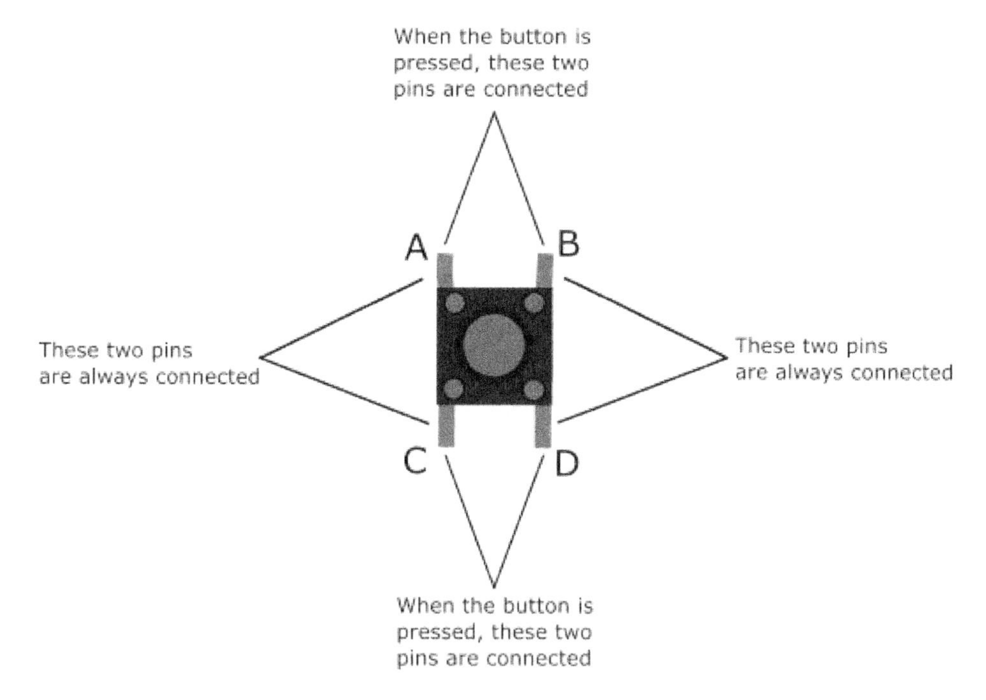

Figure 82 - The internal connections in a push button

When the button is pressed, current is able to flow from pin B to pin A, and then down to pin C.

Things to Try

Try increasing the size of R2 to see what happens. LED2 should get dimmer, and when the resistance value is sufficiently big, there won't be enough current at the Base of the transistor to switch on the transistor. At this point, current will no longer flow from the Collector to the Emitter.

Try redesigning the circuit to use a PNP transistor instead of an NPN transistor. If you need help on how to do this, go back and have a look through Chapter 8 on Page 71.

YouTube Video

There is a YouTube video to accompany this chapter. You can find it here:

https://youtu.be/8ZEQEV-Stkc

Chapter 20

Charging and Discharging a Capacitor

The purpose of this experiment is to design and build a circuit to charge and discharge a capacitor.

Parts Needed

	Breadboard, 9 V Battery and Battery Clip or Box As for previous experiments.
	Resistors We need at least two resistors for this circuit: one (470 Ω) to protect the LED and a second one to be connected in series with the capacitor (see below). The value of the second resistor needs to be something in the region of 1 kΩ or 10 kΩ.
	Capacitors We need at least one capacitor. If you have a range of sizes so much the better. Capacitors in the range 100 μF to 1000 μF will be just right.
	Push Buttons This circuit requires two push buttons - one is used to charge the capacitor and one is used to discharge it.
	Diode (optional) Although the circuit will work without using a diode, I've included one to show how it can be inserted into a circuit to only allow current to flow in a particular direction. Use a rectifier diode like the one shown here, which has a grey band to indicate which pin is the cathode (for

	example a 1N4001).
	Jumper Wires To build the circuit you need to use three jumper wires. The colour or length doesn't matter, but short wires will be more practical.
	LED We use one LED, which is briefly illuminated by the charged capacitor when the discharge button is pressed.
	Multimeter (optional) If you have a multimeter you can connect it to the capacitor to see the voltage increase and decrease as the capacitor is charged and discharged.
	Oscilloscope (optional) If you have an oscilloscope you can connect it to the capacitor to see the waveform produced when the capacitor is charged and discharged.

Table 24 - Parts needed for a capacitor charge/discharge circuit

Designing the Circuit

Figure 83 shows a schematic for the circuit.

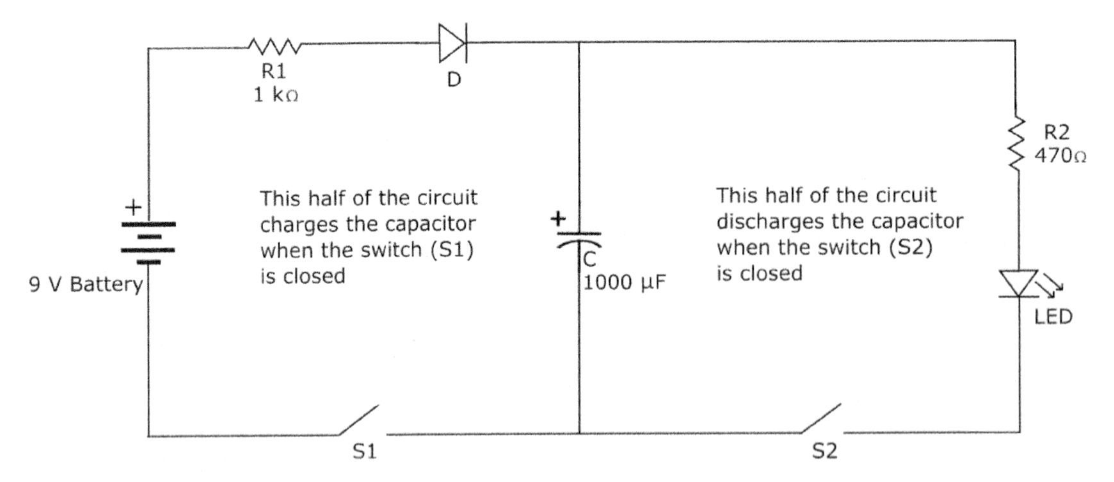

Figure 83 - Charging and discharging a capacitor

The circuit can be thought of as having two halves: one half (the left-hand side) to charge the capacitor and the other half (the right-hand side) to discharge it. So let's examine each of these halves to see how the circuit as a whole works.

Charging the Capacitor

This is the left-hand half of the circuit shown in *Figure 83* and comprises:

- a 9 V battery which is used to charge the capacitor,
- a 1 kΩ resistor (R1) which influences how long it takes to charge the capacitor,
- a diode which only lets current flow in one direction (see the explanation of the second half of the circuit),
- the 1000 μF capacitor (C) itself which, once charged, can be used to power the LED, and
- a switch (S1) which is used to complete the circuit in this half of the circuit.

When the switch (S1) is closed, electric current is able to flow around this half of the circuit. (*The current can't flow around the right-hand half of the circuit because the second switch (S2) is open.*) The current flows from the positive terminal on the battery, through the resistor, diode, capacitor and switch, before returning to the negative terminal.

Connecting the resistor (R1) and capacitor (C) in series forms what is known as a resistor-capacitor (or RC) circuit, which is a very common circuit used in electronics, and which we discussed briefly in Chapter 10 on Page 80 when we examined the 555 Timer. The values of these two components - in ohms for the resistor and farads for the capacitor - determine how long it takes for the capacitor to charge. This is known as the time constant (TC) and is calculated using the following formula:

$T = R \times C$

where:

- T = time constant (in seconds)

- R = value of resistance (in ohms)
- C = value of capacitance (in farads)

Example 1

To calculate the time constant for a 1 kΩ resistor and 1000 μF capacitor, firstly we need to convert the resistance value into ohms, and the capacitance value into farads:

1 kΩ = 1000 Ω

1000 μF = 0.001 F

Therefore, to calculate the time constant (T):

T = R x C

T = 1000 x 0.001

T = 1 second

So, we have calculated that it takes 1 second to charge the capacitor. In fact, this isn't quite true because after one time constant the capacitor is only about 66% charged. It will take another four time constants before we can confidently say that the capacitor is fully charged. So now, we can see that instead of it taking 1 second to charge the capacitor, it will in fact take about 5 seconds.

Example 2

If we increase the size of the resistor to, say, 10 kΩ, this will have a big impact on how long it takes to charge the capacitor:

10 kΩ = 10,000 Ω

Therefore, to calculate the time constant (T):

T = R x C

T = 10000 x 0.001

T = 10 seconds

As we mentioned above, it takes five time constants for the capacitor to be fully charged, so to be fully charged it will take in the region of 50 seconds.

Example 3

If we change the size of the capacitor to 100 μF rather than 1000 μF, this will also have an impact on the time constant. Keeping the resistor at 1 kΩ, we now have:

T = R x C

T = 1000 x 0.0001

T = 0.1 seconds

Therefore, to be fully charged it will take about 0.5 seconds ... so pretty quick.

Sample Oscilloscope Trace for Our 1 kΩ / 1000 μF

To make it easier to visualize what's going on, look at the following figure.

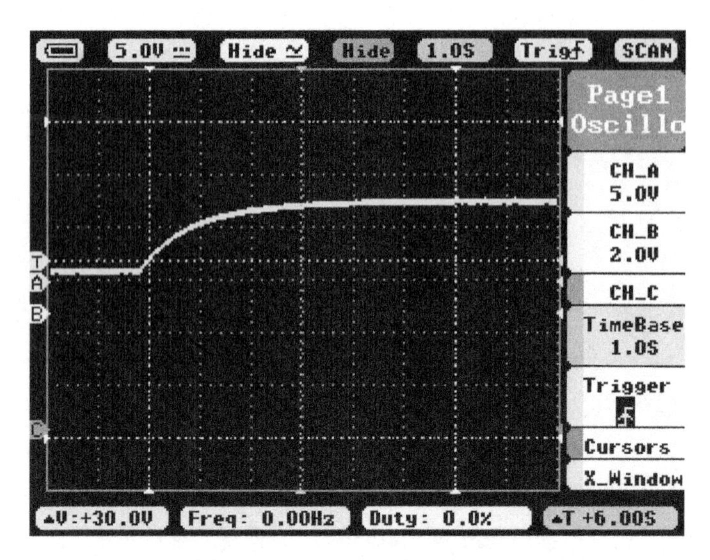

Figure 84 - Oscilloscope trace when charging a 1000 μF capacitor in series with a 1 kΩ resistor

Each of the small squares represents 1 second. Initially, the capacitor charges very quickly (about 60-70% in the first second or so), but then it takes a lot longer to become fully charged (roughly another 4 seconds). So, I think, the oscilloscope scan supports the time constant theory we discussed above.

Discharging the Capacitor

It is the right-hand side of the circuit shown in *Figure 83* on Page 132 that discharges the capacitor. It comprises:

- the capacitor,
- a 470 Ω resistor (R2) which is used to protect the LED (remember Ohm's Law from earlier in the book),
- the LED, and
- a second switch (S2).

When the switch (S2) is closed, current flows from the capacitor's positive terminal, through the resistor (R2) and the LED before arriving back at the capacitor's negative terminal.

If you remember when we looked at the left-hand side of the circuit I mentioned the diode. What this does is to prevent the current from trying to flow around the left-hand side of the circuit (towards the battery's positive terminal) when the switch (S2) is closed.

As it happens, the circuit would work perfectly well without the diode being present, but it is a good idea to get into the habit of using diodes to ensure that current only flows in the direction in which you want it to flow. For example, in some circuits there may well be components that would get damaged if current hits them in the wrong direction.

The amount of time it takes for the capacitor to discharge can be calculated by using the same formula as we used above to calculate the capacitor charge time, in other words:

T = R x C

So, in our RC circuit this time - the right-hand side of the circuit shown in *Figure 83* on Page 132 - we have the 1000 μF capacitor and a 470 Ω resistor.

Therefore, the time to discharge the capacitor is:

T = 470 x 0.001

T = 0.47 seconds

This is the time to discharge the capacitor by about 60-70%. We need to multiple this value by 5 to calculate the time it takes to fully discharge the capacitor... 0.47 x 5 = 2.35 seconds.

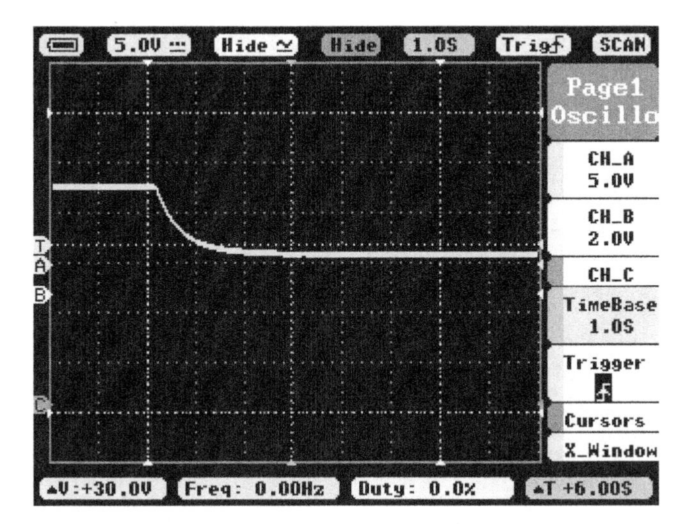

Figure 85 - Oscilloscope trace when discharging a 1000 μF capacitor in series with a 470 Ω resistor

As before, the small squares represent 1 second, so we can see that after about 2.5 seconds the capacitor is pretty much fully discharged.

Building the Circuit on a Breadboard

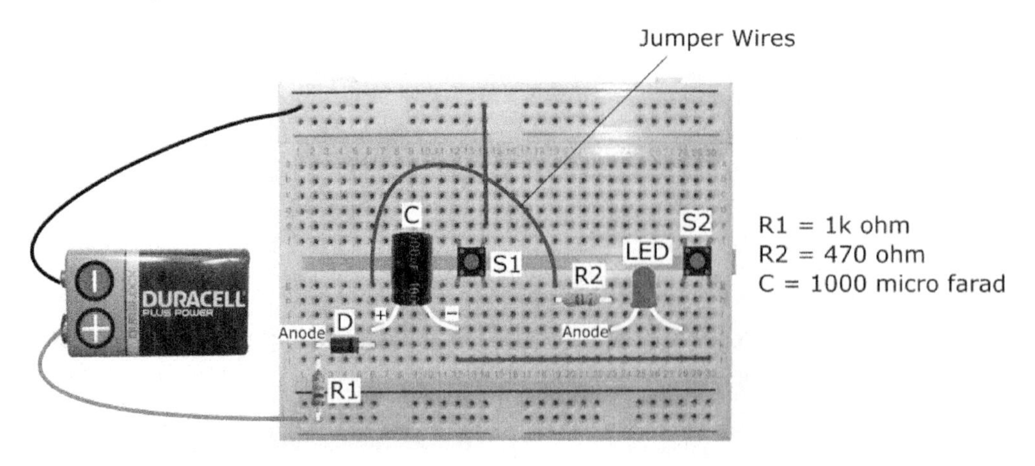

Figure 86 - Building the circuit on a breadboard

What Happens

Refer to the text earlier in this chapter for an explanation of how the circuit works. Basically though, when you press S1, the capacitor charges, and when you press S2 it discharges, which illuminates the LED for a short period of time.

Things to Try

If you have a multimeter you can connect it to the two pins on the capacitor to see the voltage increase and decrease across the pins as the capacitor charges and discharges. If you have an oscilloscope, you can view the trace produced as the capacitor charges and discharges. I'll look at both of these things in the video associated with this chapter (see below).

If you have a selection of capacitors and resistors of different sizes, try a variety out to see what impact they have on the time it takes to charge and discharge the capacitor. Remember though, the LED needs to be protected, so don't go much smaller than 470 Ω for R2.

Finally, try pressing both buttons S1 and S2 at the same time to see what happens. The LED will be permanently illuminated ... but why? ... I'll leave you to figure that one out.

YouTube Video

There is a YouTube video to accompany this chapter. You can find it here:

https://youtu.be/u_l-Mz5zDiM

Chapter 21

Building a Light Sensor - Night Light

The purpose of this experiment is to show how a Light Dependent Resistor (LDR), a variable resistor and a PNP transistor can be used together to create a night light circuit. This is a circuit whereby an LED is illuminated when it gets dark. At the end of the chapter we also examine how to turn the circuit into a day light.

Parts Needed

	Breadboard, 9 V Battery and Battery Clip or Box As for previous experiments.
	470 Ω Resistor As usual, we use a 470 Ω resistor to limit the amount of current flowing through the LED.
	PNP Transistor We need one PNP transistor. I use a BC 556 B BJT PNP but you could use a different one if you prefer. Make sure you know which pins are designated as the Base, Collector, and Emitter though as this is essential when it comes to building the circuit.
	Jumper Wires We need a few jumper wires. The colour or length doesn't matter, but short wires will be more practical.
	LED We use one LED. Any colour will do.

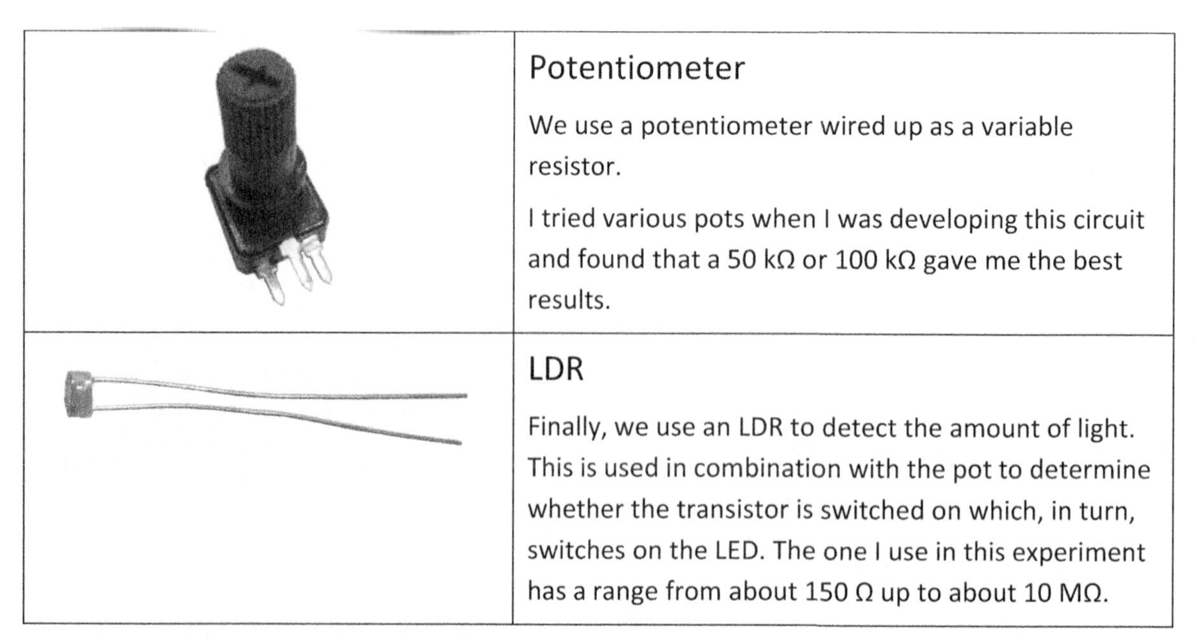

	Potentiometer We use a potentiometer wired up as a variable resistor. I tried various pots when I was developing this circuit and found that a 50 kΩ or 100 kΩ gave me the best results.
	LDR Finally, we use an LDR to detect the amount of light. This is used in combination with the pot to determine whether the transistor is switched on which, in turn, switches on the LED. The one I use in this experiment has a range from about 150 Ω up to about 10 MΩ.

Table 25 - Parts needed for the circuit

Designing the Circuit

Figure 87 shows a schematic for the circuit.

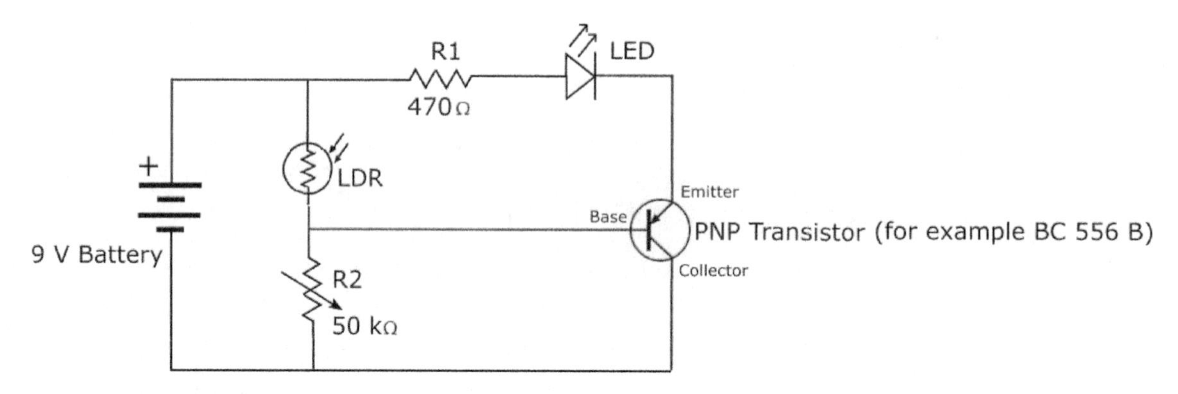

Figure 87 – Night light circuit diagram

Basic Circuit Operation

Before reading this section you might find it useful to refresh your memory with regard to how a PNP transistor works by re-reading some or all of Chapter 8 on Page 71.

The three interesting components in the circuit are the transistor, which behaves like an on/off switch; the LDR, which determines the voltage applied to the base of the transistor, which either switches it on or off – by controlling the flow of current from the Emitter to the Collector; and the variable resistor, which - working in conjunction with the LDR - determines the level of light at which the transistor - and therefore the LED - is switched on.

The resistance of the LDR changes according to how much light it is exposed to:

- When it is dark (night time), the resistance is high
- When it is light (day time), the resistance is low

This change in resistance determines whether the transistor is switched on or not.

The variable resistor works in conjunction with the LDR to control the level of light at which the transistor is switched on. You could use a fixed 50 kΩ (or 100 kΩ) resistor instead, but you would have no adjustment in the circuit and you might find that the LED is coming on when you don't want it to, or not coming on when you do want it to.

When I was designing the circuit I had a handful of variable resistors and I played around with them until I achieved an acceptable result. I could have done the same thing with a load of fixed-value resistors but it may have taken me longer to get there. If, for example, you find that a 50 kΩ doesn't work for you, try a 20 kΩ or a 100 kΩ, and adjust the resistance value using the knob until you get to a situation whereby the LED comes on when it is dark. For the circuit shown in *Figure 87* I use a 50 kΩ variable set to the maximum resistance.

Daytime Operation

When it is light during the day, the resistance of the LDR is low – much lower than R2. This means that the voltage being supplied to the Base of the transmitter is high (close to 9 V). If the voltage at the Base is higher than the voltage at the Emitter, the transistor is not switched on, so the LED is not illuminated.

Night time Operation

When it is dark at night, the resistance of the LDR is high – much higher than R2. This means that the voltage being supplied to the Base of the transmitter is *more negative* than that being supplied to the Emitter. This switches on the transistor and current flows between the Emitter and Collector, which completes the circuit and illuminates the LED.

Building the Circuit on a Breadboard

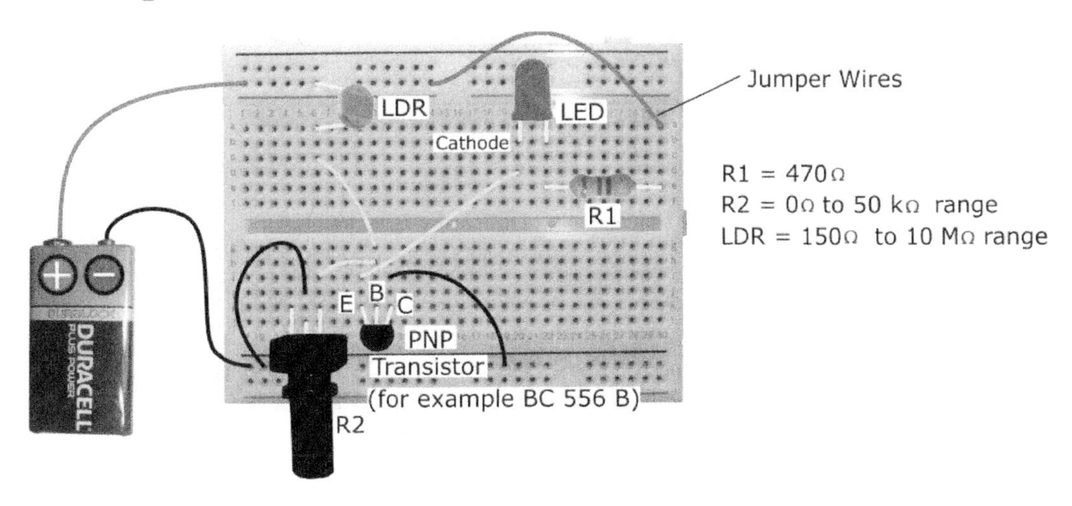

Figure 88 - Building the night light circuit on a breadboard

What Happens

If you build the circuit in a reasonable amount of light, which I expect you will otherwise you won't be able to see what you're doing, when you switch on the circuit it's quite possible that the LED will not be illuminated. Assuming you have built the circuit correctly, this is the correct behaviour. If you now cover the LDR with your fingers to simulate darkness, the LED should become illuminated. If it doesn't, turn the knob on the variable resistor to change (reduce) the resistance (if you don't know which way to turn it – try both directions). At some point, the LED should become illuminated. This is because the voltage at the Base of the transistor is more negative than at the Emitter, which has switched on the transistor (causing current to flow from the Emitter to the Collector), which in turn illuminates the LED.

If, on the other hand, the LED is already illuminated when you switch on the power, try turning the knob on the variable resistor until the LED switches off.

The variable resistor enables you to set the sensitivity of the night light so that you can determine at what level of light the LED becomes illuminated. Try rotating the knob on the variable resistor in both directions while covering (or partially covering) the LDR. By doing this you should be able to set the circuit up (calibrate it) so that the LED comes on when you want it to.

Things to Try

If you have any other variable resistors with different resistance values to the one you used above, try them out in the circuit to see what impact they have. If you haven't got any variable resistors, just use fixed value ones.

What you should find is that if you use a resistor with a resistance value that is quite low, say 1 kΩ, the LED is on all the time, even in daylight. This is because the transistor is permanently switched on. If, on the other hand, you go for a very high resistance, say 1 MΩ, you might find that the LED won't come on at all - even when it's very dark.

The key is to play around with various resistors until you find one that suits your needs. If you have a few LDRs, you could also try using different ones to see what impact they have.

Day Light

It's very simple to change the circuit so that it operates as a day light rather than a night light. In fact, all you need to do is to replace the PNP transistor with an NPN transistor and swap the connections for the Emitter and Collector. I also found that I needed to replace the 50 kΩ variable resistor with a 100 kΩ one.

Here are the schematic and breadboard diagrams.

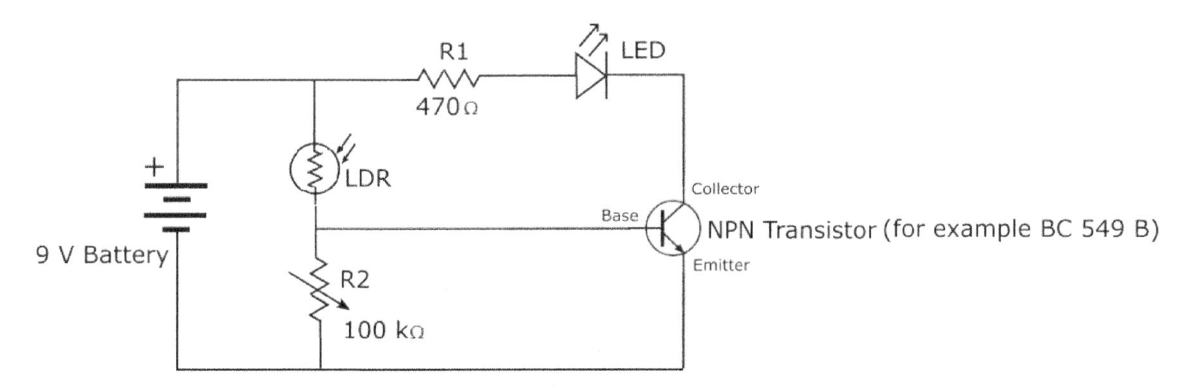

Figure 89 – Day light circuit diagram

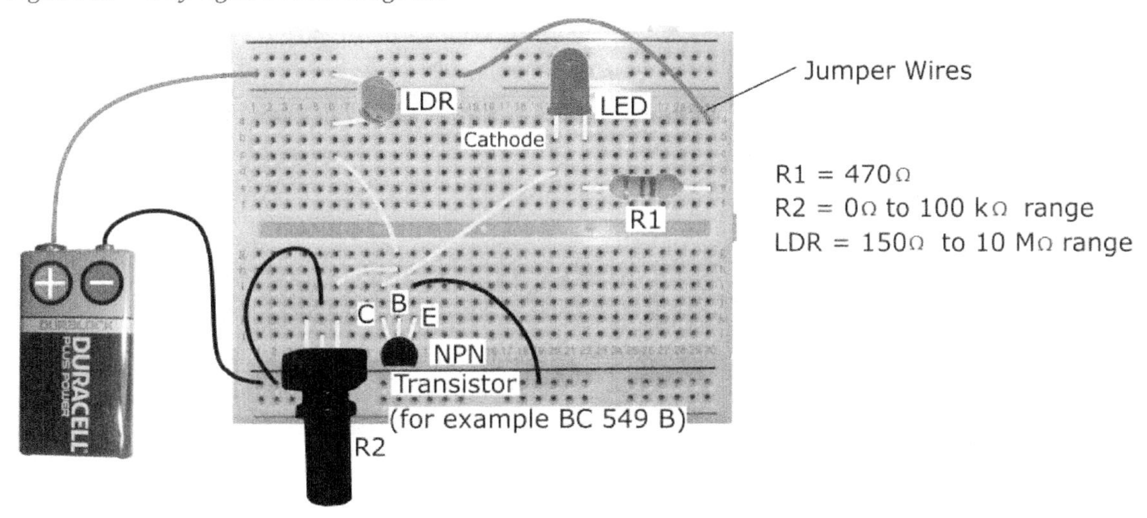

Figure 90 - Day light breadboard circuit

> *Note*
>
> *Note that although all the jumper wires are in the same place as they were in Figure 88* on Page 139, *the pin positions are different for the transistor.*

When it's light, the resistance of the LDR is low, which allows enough current to reach the Base to switch on the transistor. This then allows current to flow from the Collector to the Emitter, which completes the circuit and illuminates the LED.

YouTube Video

There is a YouTube video to accompany this chapter. You can find it here:

https://www.youtube.com/watch?v=-yg9h1GxZVw

Chapter 22

Introduction to Solar Power

In this chapter we discuss how to use solar panels to harness the free energy provided by the Sun, and in doing so create a solar-powered version of the day light circuit created at the end of the previous chapter (see *Figures 89* and *90*).

> *Why a daylight and not a (more useful) night light?*
>
> *When I first started thinking about this chapter, my intention was to use a solar panel to charge some rechargeable batteries during the day, which in turn would power the night light when it was dark. The circuit worked as planned, but the problem was that it did not safeguard the batteries from potentially overcharging.*
>
> *As I have no control over the type of rechargeable batteries someone might use, or how much bright sunlight the solar panel might be exposed to, I decided to opt for the safe option and just demonstrate how a solar panel can be used to directly power a circuit ... without the use of batteries.*

Solar Cells and Panels

A solar panel is made up of an array of solar cells (also called photovoltaic cells), which generate electric current when exposed to light.

Solar cells consist of two silicon semiconductor layers sandwiched between a solid conductor on one side (the back) and gridded connectors on the other (the front). Light is able to pass through the grid to the silicon layers. One silicon layer is an N-Type, which has an excess of electrons, while the other is a P-Type, which has a deficiency of electrons (electron holes).

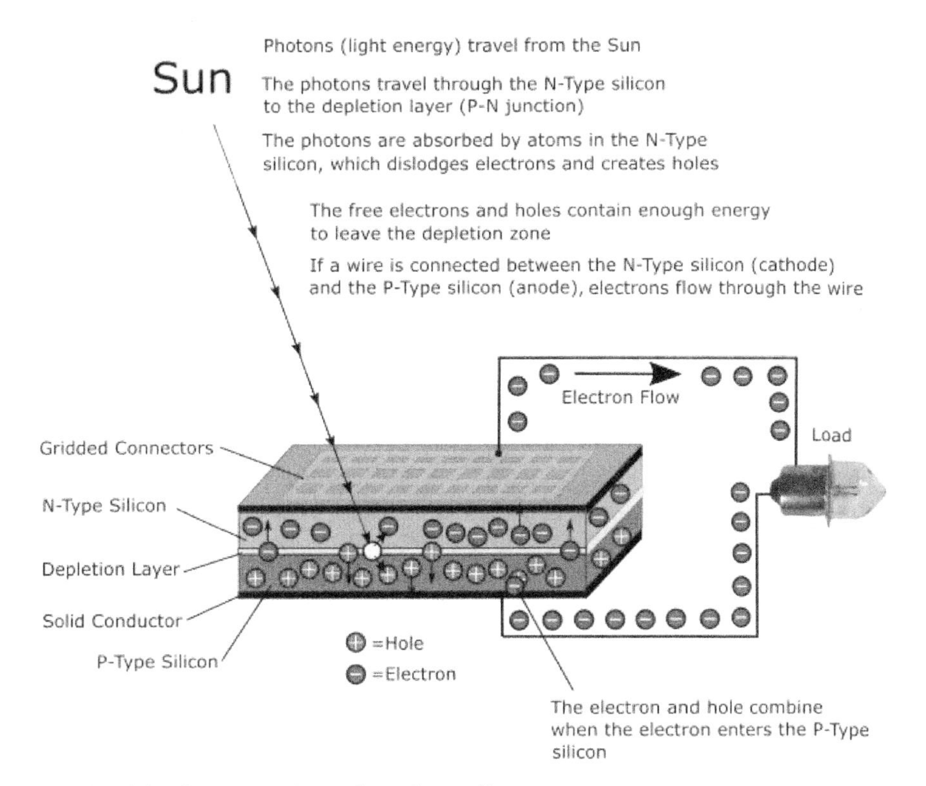

Figure 91 - Construction of a solar cell

When photons of light hit the solar cell, electrons are knocked out of position at the P-N junction (depletion layer). When a wire is connected to the cell's conductors, the electrons leave the negative layer and travel through the wire to the electron holes on the positive layer.

The more cells a solar panel has, the higher the voltage that can be generated; the larger the cells, the more current that can be generated.

Solar panels are normally used to charge batteries rather than being used to directly power electrical equipment.

Increasing the current produced by solar panels

You can increase the amount of current produced by connecting multiple solar panels in parallel, as shown below in *Figure 92*.

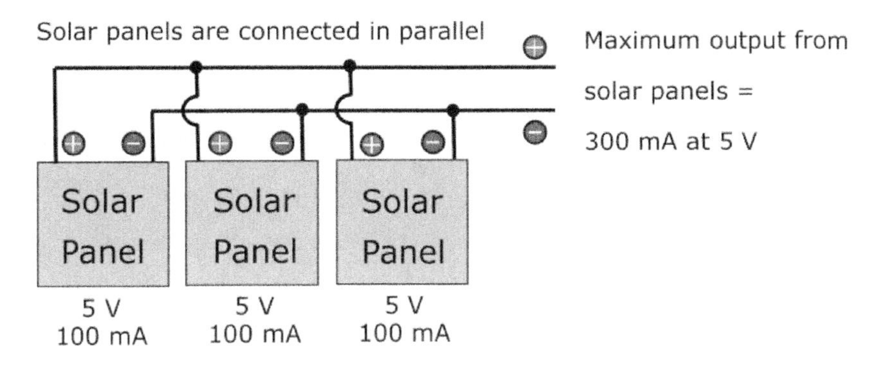

Figure 92 – Connecting solar panels in parallel

Increasing the voltage produced by solar panels

You can increase the amount of voltage produced by connecting multiple solar panels in series, as shown below in *Figure 93*.

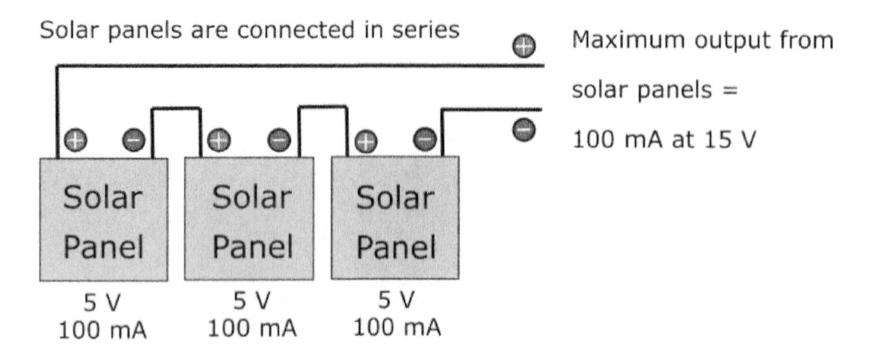

Figure 93 – Connecting solar panels in series

Parts Needed

Most of the parts required for this chapter's experiment are the same as those used in the night light circuit from the previous chapter. For completeness though, I've included them in the following table, plus any new components required for this chapter's experiment.

	Breadboard ... as usual.
	Solar Panel In order to get our free solar energy from the Sun, we need to use a small solar panel. The one I use in this experiment can produce a maximum output of 5 V and 100 mA, giving a maximum power output of 0.5 W.
	Resistor As usual, in order to protect the LED we need to use a resistor. This experiment is a little different to previous ones though because the voltage is not fixed. When we used batteries we knew what the voltage value was, but with a solar panel the voltage will vary depending on how sunny it is.

	The maximum voltage produce by my solar panel is 5 V, although in general it will most likely be much less than this. For my version of the circuit, I'll use a resistor of about 80 Ω. If this turns out to be too small – in other words the LED goes pop – I'll use a bigger one.
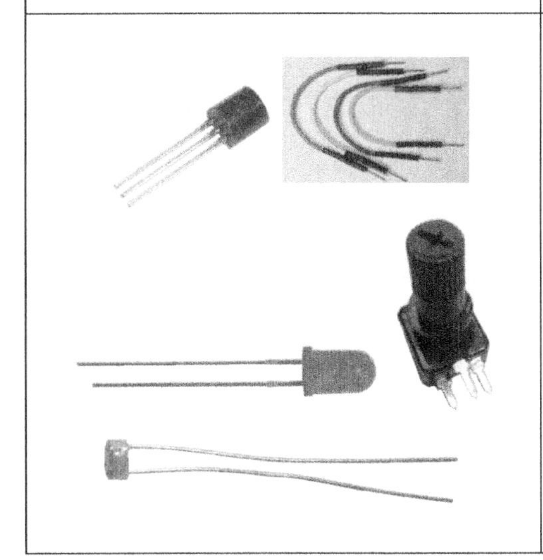	**Other Stuff** We need one NPN transistor (for example, a BC 549 B), some jumper wires, an LED, a potentiometer (50 kΩ or 100 kΩ), and an LDR.

Table 26 - Parts needed for the circuit

Difference between electrical energy and power

Electric power *is the rate, per unit time, at which* ***electrical energy*** *is transferred by an* ***electric*** *circuit. The SI unit of* ***power*** *is the watt, one joule per second.* ***Electric power*** *is usually produced by* ***electric*** *generators, but can also be supplied by sources such as* ***electric*** *batteries.*

Source: *https://en.wikipedia.org/wiki/Electric_power*

Designing the Circuit

Here are the schematic and breadboard diagrams for the circuit.

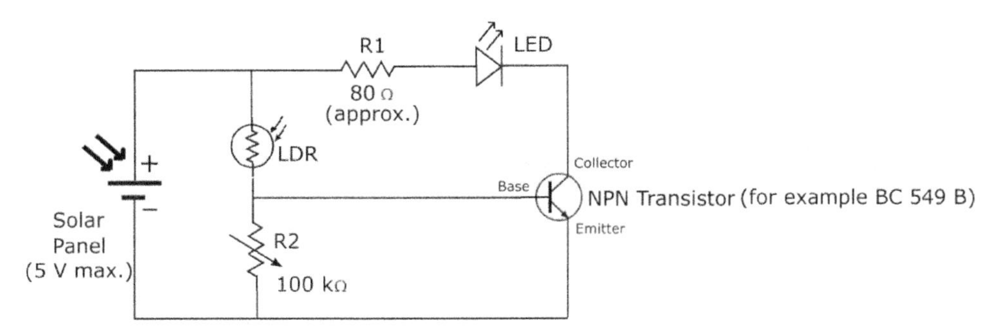

Figure 94 – Solar powered day light circuit diagram

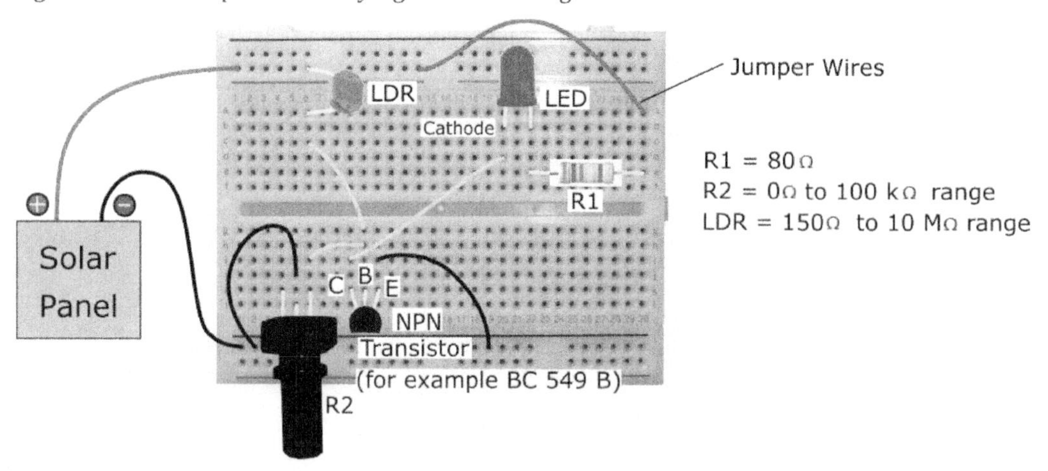

Figure 95 - Day light breadboard circuit

Being as the amount of voltage output from the solar panel will vary, you may find that you need to use a different size resistor for R1 ... or perhaps you don't need to use one at all.

Background Information: Using a Solar Panel to Charge Batteries

Although – as stated earlier – we won't be using rechargeable batteries in this experiment, here is a bit of information about how long it should typically take to charge and discharge them.

One of the specifications provided with rechargeable batteries is the mAh value, for example, 1200 mAh. What this means is that if you attach a load (to the batteries) that requires 100 mA to operate, the batteries will (theoretically) last 12 hours (assuming they are fully charged) before they become flat. Likewise, if you attach a load that requires 200 mA to operate, the batteries will last 6 hours before they become flat.

If you use a solar panel that can produce 100 mA in full sunlight, it will take 12 hours to (theoretically) fully charge the batteries. Of course, any inefficiencies in the circuit - along with a lack of bright sunlight - will have an impact on how long it really takes to charge the batteries. For example, my solar panel can in theory provide 100 mA, but the reality is that it has never provided more than about 60 mA, even on a very sunny (United Kingdom) day. The following table shows the approximate current and voltages my solar panel supplies for various weather conditions (in the UK).

Weather Condition	Current	Voltage
Sunny Day	30 mA to 60 mA	4.5 V to 5 V
Hazy	10 mA to 20 mA	3.5 V to 4.5 V
Dull	50 µA to 5 mA	1.5 V to 2.0 V

Rainy	< 20 µA	< 1.5 V

Table 27 – Approximate current and voltage values for my 100 mA, 5 V solar panel

You can estimate how long it will take to fully charge a rechargeable battery by using the following formula:

(Capacity of Battery / Charge Rate) + 20%

You can find a calculator, plus lots of useful information, here:

http://convert-to.com/491/recharging-rechargeable-nimh-nicd-batteries.html

YouTube Video

There is a YouTube video to accompany this chapter. You can find it here:

https://youtu.be/WOnrJ9MICg4

Printed in Great Britain
by Amazon